Aa

ab·do·men (ab′ də mən) *noun* 1. In humans, the part of the body between the chest and the hips. *Your stomach is located in your abdomen.* 2. One of the three parts of an insect's body. *An insect has a head, thorax, and abdomen.*

ab·sorb (əb zôrb′) *verb* To soak up or take in. *Sponges absorb liquids.*

ab·sorp·tion (əb zôrp′ shən) *noun* The process of soaking up liquid or taking in heat. *Roots are parts of a plant where absorption takes place.*

a·buse 1. (ə būz′) *verb* To hurt or treat cruelly. *Pet owners should never abuse their animals.* 2. (ə būs′) *noun* Cruel treatment or harmful misuse. *Drug abuse is dangerous.*

ac·cel·er·ate (ak sel′ ər āt) *verb* To go faster. *Roller coasters accelerate when they go downhill.* **acceleration** *noun*

ac·id (as′ id) *noun* A substance that tastes sour, turns blue litmus paper red, and reacts with a base to form water and a salt. *Strong acid can burn your skin.*

ac·id rain (as′ id rān) *noun* Rain that is polluted by wastes that got into the air from burned fuels. *Acid rain is harmful to lakes, crops, and even buildings.*

ad·ap·ta·tion (ad əp tā′ shən) *noun* A body part or behavior that helps a living thing survive in an environment. *A porcupine's quills are an adaptation that keeps it safe from its enemies.* **adapt** *verb*

ad·he·sion (ad hē′ zhən) *noun* The attraction between particles of different substances. *Adhesion causes raindrops to stick to windowpanes instead of bouncing off.* **adhesive** *adjective*

a·dult (ə dult′) *noun* An organism that is fully developed. *A butterfly is an adult insect.*

air (âr) *noun* The mixture of gases all around Earth. *Our bodies get oxygen from the air we breathe.*

air mass (âr mas) *noun* A large body of air of nearly the same temperature and humidity. *The cold air mass brought freezing temperatures to the city.*

air pol·lu·tion (âr pə lü′ shən) *noun* The process of putting wastes or poisonous substances into the atmosphere. *Automobiles are a major source of air pollution.*

air pres·sure (âr presh′ ər) *noun* The force of air pressing down on Earth in an area. *Air pressure usually goes down when a storm is coming.*

air re·sis·tance (âr ri zis′ təns) *noun* The force felt by an object moving through air as the object hits air molecules. *The curved front of a race car cuts down air resistance.*

air sac (âr sak) *noun* A baglike part of the lung that holds air; also called alveoli. *Air sacs are grouped like bunches of balloons at the ends of airways in the lungs.*

al·co·hol (al′ kə hôl) *noun* A colorless liquid used in making chemicals and fuels. *Alcohol is used in many medicines.*

al·gae (al′ jē) *noun, plural* Plantlike organisms, often one-celled, that grow in water. *Algae have no stems, leaves, roots, or flowers.* **alga** *singular*

al·ler·gy (al′ ər jē) *noun* The body's reaction to a substance that is not harmful to most people. *Sneezing and coughing are common signs of allergy.* **allergic** *adjective*

al·loy (al′ oi) *noun* A mixture of two or more metals. *Brass is an alloy of copper and zinc.*

al·ter·na·ting cur·rent (ôl′ tər nāt iŋ kûr′ ənt) *noun* Electric current that flows first in one direction and then in the opposite direction, changing at regular time periods. *Most homes are wired for alternating current, or AC.*

al·ti·tude (al′ ti tüd) *noun* 1. The height of an object above Earth's surface. *Cirrus clouds have an altitude of at least 6,000 meters (20,000 feet).* 2. The angle between an object in the sky and the horizon. *As the Moon rises, its altitude gets larger.*

a·lu·mi·num (ə lü′ mə nəm) *noun* An element that is a lightweight, silver-colored metal. *Many soft drinks come in aluminum cans.*

al·ve·o·li (al vē ō′ lī) *noun, plural* Tiny air bags in the lungs; also called air sacs. *Oxygen passes through the walls of alveoli into the bloodstream.* **alveolus** *singular*

am·ber (am′ bər) *noun* Hardened sap of certain trees. *We can find insect fossils in amber.*

am·phib·i·an (am fib′ ē ən) *noun* A cold-blooded animal that lives both on land and in the water. *Amphibians such as frogs begin life in the water.*

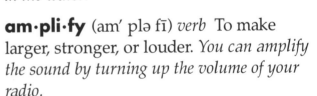

am·pli·fy (am′ plə fī) *verb* To make larger, stronger, or louder. *You can amplify the sound by turning up the volume of your radio.*

a·ne·mi·a (ə nē′ mē ə) *noun* An unhealthy lack of red blood cells or of iron in the red blood cells. *A person with anemia is often tired and pale.*

a	cat	e	net	îr	gear	u	cup	u̇	look, pull	*th*	**this**	ə	alive,
ā	day, lake	ē	seed	o	hot	ū	fuse	oi	soil	hw	**wheel**		comet,
ä	father	i	fit	ō	cold	ûr	fur, bird	ou	out	zh	measure		acid, atom,
âr	dare	ī	pine	ô	paw	ü	tool, rule	th	thin	ŋ	wing		focus

an·e·mom·e·ter (an ə mom' i tər) *noun* An instrument for measuring wind speed. *The anemometer was spinning fast during the storm.*

an·i·mal (an' ə məl) *noun* A living thing that has many cells, can move, reacts to things around it, and eats plants or other animals to live. *Animals, such as fish, birds, insects, and mammals, make up one of the five kingdoms of living things.*

an·ten·na (an ten' ə) *noun* Sense organs attached to an insect's head, also called feelers. *Insects use their antennae to hear, taste, feel, and smell.*
antennae, antennas *plural*

an·ti·bi·ot·ic (an' tē bī ot' ik) *noun* A medicine that kills or slows the growth of germs that cause disease. *Penicillin is an important antibiotic.*

an·ti·bod·y (an' tē bo dē) *noun* A substance the body makes to fight infection. *White blood cells carry antibodies that keep us healthy.*

an·ti·sep·tic (an ti sep' tik) *noun* A substance that prevents infection by slowing or stopping the growth of germs. *Put an antiseptic on your cut.*

a·or·ta (ā ôr' tə) *noun* The main artery in the body. *The aorta carries blood from the heart to the rest of the body.*

ar·ea (âr' ē ə) *noun* The size of a surface, measured in square units. *The area of the desktop is 6 square feet.*

ar·ter·y (ärt' ər ē) *noun* A blood vessel that carries blood away from the heart. *The blood in arteries is rich in oxygen.*

ar·thro·pod (är' thrə pod) *noun* An animal with jointed legs and a body divided into sections. *Insects, spiders, crabs, and lobsters are arthropods.*

as·ter·oid (as' tər oid) *noun* A piece of rock that travels around the Sun. *Most asteroids orbit in an area between Mars and Jupiter.*

as·tron·o·my (ə stron' ə mē) *noun* The scientific study of objects and events in space. *Stars, planets, moons, and galaxies are all studied as part of astronomy.*

at·mo·sphere (at' mə sfîr) *noun* The blanket of gases that surrounds a planet or moon. *Weather occurs in the layer of the atmosphere closest to Earth.*

at·om (at' əm) *noun* The smallest particle of an element. *Atoms are the basic building blocks of all matter.*

a·tom·ic fis·sion (ə tom' ək fish' ən) *noun* The process of splitting atoms. *Atomic fission can release large amounts of energy.*

a·tri·um (ā' tri əm) *noun* One of the two upper chambers of the heart, into which blood flows. *The right atrium collects blood carried by the veins from the body.*
atria, atriums *plural*

at·tract (ə trakt') *verb* To pull toward. *Opposite poles of magnets attract each other.*
attraction *noun*

au·di·to·ry nerve (ô' də tôr ē nûrv) *noun* The nerve that carries messages from the ear to the brain. *The brain changes the messages carried by the auditory nerve into sounds.*

ax·is (ak′ sis) *noun* The imaginary straight line around which a planet rotates. *Earth rotates on its axis from west to east one full turn every 24 hours.*

back·bone (bak′ bōn) *noun* The long row of connected bones that runs down the back of humans and many other animals; also called spine. *Animals with backbones are called vertebrates.*

bac·ter·ia (bak tîr′ ē ə) *noun, plural* One-celled organisms that exist all around you and inside you. *Some bacteria can cause diseases.* **bacterium** *singular*

bal·ance (bal′ əns) *noun* An instrument for measuring mass. *The equal-arm balance is useful for comparing the masses of two objects.*

bal·anced (bal′ ənst) *adjective* Set up so that parts are equal in weight, value, or importance. *A balanced diet includes all the food groups.*

ball-and-soc·ket joint (bôl ənd sok′ it joint) *noun* A joint in the body that allows bones to swing or rotate in a full circle. *Shoulders and hips are ball-and-socket joints.*

bar graph (bär graf) *noun* A graph that compares data using bars of different heights. *Draw a bar graph showing the average rainfall for each month.*

ba·rom·e·ter (bə rom′ i tər) *noun* An instrument for measuring air pressure. *High readings on a barometer are a sign of fair weather.* **barometric** *adjective*

bar·ri·er is·land (bâr′ ē ər ī′ lənd) *noun* A thin island offshore along a coast. *Waves crashed over the barrier islands during the hurricane.*

ba·salt (bə sôlt′) *noun* A dense, dark igneous rock that has small grains. *The crust on the ocean floor is mostly basalt.*

base (bās) *noun* A substance that tastes bitter and reacts with an acid to form water and a salt. *Bases turn red litmus paper blue.*

a	cat	e	net	îr	gear	u	cup	u̇ look, pull
ā	day, lake	ē	seed	o	hot	ū	fuse	oi soil
ä	father	i	fit	ō	cold	ûr	fur, bird	ou out
âr	dare	ī	pine	ô	paw	ü	tool, rule	th thin

th	**this**	ə alive,
hw	**wheel**	comet,
zh	measure	acid, atom,
ŋ	wing	focus

bat·ter·y (bat' ə rē) *noun* A container that holds chemicals to produce electric current. *The car needs a new battery.*

bay (bā) *noun* A portion of the ocean that is partly enclosed on two sides by land. *When the storm came, the boats sailed into the bay.*

Beau·fort scale (bō' fûrt skāl) *noun* A method for estimating wind speed by observing the wind's effect on objects such as trees. *The Beaufort scale rates the force of wind from 0 (calm) to 12 (hurricane strength).*

bed·rock (bed' rok) *noun* The solid layer of rock under the soil. *Wells are sometimes drilled through bedrock.*

be·ha·vior (bi hāv' yər) *noun* The way a person or thing acts. *Weather scientists study the behavior of hurricanes.*
behave *verb*

Ber·noul·li ef·fect (bûr nü' lē ə fekt') *noun* The principle that fast-moving air has less pressure than slow-moving air. *Airplanes lift off the ground and fly partly because of the Bernoulli effect on the tops of the wings.*

bi·ceps (bī' seps) *noun, singular, plural* The muscles in the front of the upper arm. *Biceps contract, or get shorter, to pull up your lower arm.*

bi·cus·pid (bī kus' pid) *noun* One of eight teeth in front of the molars. *Bicuspids are used to grind food.*

bi·lat·er·al sym·me·try (bī lat' ər əl sim' ə trē) *noun* A shape or design with matching parts on both sides of a center line. *Butterflies have bilateral symmetry.*

bi·o·sphere (bī' ə sfîr) *noun* The parts of the land, water, and air where living things can be found. *The biosphere includes all of nature.*

bi·ot·ic po·ten·tial (bī ot' ik pō ten' shəl) *noun* The largest possible increase in a population. *The biotic potential of a group of animals is high if the animals have few enemies.*

bird (bûrd) *noun* A warm-blooded animal that has two legs, wings, feathers, and a beak. *Birds lay eggs, and most birds can fly.*

blad·der (blad' ər) *noun* A baglike organ in which liquid wastes are collected before they leave the body. *Wastes pass from the kidneys to the bladder.*

blood (blud) *noun* The red liquid that the heart pumps throughout the body. *Blood carries food and oxygen to all the organ systems.*

blood clot (blud klot) *noun* An area of blood that has thickened or become a solid lump. *A blood clot can stop the flow of blood.*

blood ves·sel (blud ves' əl) *noun* A tube in the body that carries blood. *Arteries and veins are blood vessels.*

boil (boil) *verb* To heat until the substance bubbles up and turns to a gas. *When water boils, it turns to water vapor.*

boil·ing point (boil' iŋ point) *noun* The temperature at which a liquid boils. *The boiling point of water is 100°C (212°F).*

bone (bōn) *noun* Part of the skeleton. *The skeleton of an adult human has 206 bones.*

brain (brān) *noun* The organ in the head that is the center of thought, memory, feelings, and body control. *The brain is the main organ of the nervous system.*

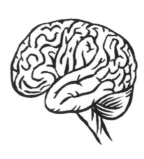

brain stem (brān stem) *noun* The base of the brain, which connects it to the spinal cord. *The brain stem controls the body's heartbeat and breathing.*

bris·tle (bris' əl) *noun* A short, stiff hair. *An earthworm moves through the soil by pulling itself forward with its bristles.*

bron·chi·al tube (bron' kē əl tüb) *noun* One of two branches of the windpipe; also called bronchus. *The left and right bronchial tubes carry the air you breathe to your lungs.*

bron·chus (bron' kəs) *noun* Another name for bronchial tube. *Each bronchus divides again and again, becoming narrower and narrower.* **bronchi** *plural*

bud·ding (bud' iŋ) *noun* A form of reproduction in which a bud forms on the parent, develops, and breaks off. *Some yeasts produce new yeasts by budding.*

buoy·an·cy (boi' ən sē) *noun* An object's ability to float. *Filling a life raft with air gives it more buoyancy.* **buoyant** *adjective*

buoy·ant force (boi' ənt fôrs) *noun* The upward push on an object by a fluid. *The bouyant force is what makes a boat float.*

burn (bûrn) *verb* To give off heat and light as part of a chemical reaction. *A fire needs oxygen to burn.*

a	cat	e	net	îr	gear	u	cup	u̇	look, pull	*th*	this	ə	alive,
ā	day, lake	ē	seed	o	hot	ū	fuse	oi	soil	hw	wheel		comet,
ä	father	i	fit	ō	cold	ûr	fur, bird	ou	out	zh	measure		acid, atom,
âr	dare	ī	pine	ô	paw	ü	tool, rule	th	thin	ŋ	wing		focus

caf·feine (kaf ēn') *noun* A substance that speeds up the activity of the heart and nervous system. *Caffeine is found in coffee, tea, and some soft drinks.*

cam·ou·flage (kam' ə fläzh) *noun* A color, shape, or pattern that helps an animal blend into its surroundings. *The speckles on a trout act as camouflage in shallow streams.*

ca·nine (kā' nīn) *noun* One of the four pointed teeth near the front of the mouth. *The canine teeth are used to tear food.*

can·yon (kan' yən) *noun* A deep, narrow river valley with steep sides. *The Grand Canyon was formed by the Colorado River.*

ca·pac·i·ty (kə pas' i tē) *noun* The amount a container can hold. *The capacity of the milk carton is 1 liter (about 1 quart).*

cap·il·lar·y (kap' ə lâr ē) *noun* A tiny blood vessel that lets oxygen and nutrients pass from blood to body cells. *Capillaries connect arteries to veins.*

car·bo·hy·drate (kär bō hī' drāt) *noun* A food substance that the body uses for fuel. *Breads, pasta, sugar, and potatoes are carbohydrates.*

car·bon (kär' bən) *noun* A chemical element found in all living things. *Carbon is also found in nonliving things, such as diamonds and coal.*

carbon 14 dat·ing meth·od (kär' bən fôr tēn' dā' tiŋ meth' əd) *noun* A process for finding the age of some fossils and other objects. *The carbon 14 dating method is used to discover the age of things up to 50,000 years old.*

carbon di·ox·ide (kär' bən dī ok' sīd) *noun* A colorless, odorless gas that animals breathe out. *Plants use carbon dioxide, water, and sunlight to produce food.*

carbon mon·ox·ide (kär' bən mə nok' sīd) *noun* A colorless, odorless, poisonous gas. *Running a car engine produces carbon monoxide.*

car·di·ac mus·cle (kär' dē āk mus' əl) *noun* The special type of muscle that forms the heart. *Cardiac muscle works all the time and never tires.*

car·ni·vore (kär' ni vôr) *noun* An animal that eats other animals for food. *Lions and tigers are carnivores.*
carnivorous *adjective*

car·ti·lage (kär' tə lij) *noun* A tough, elastic tissue that connects to bones and is part of the skeleton. *The outside part of the ear is made of cartilage.*
cartilaginous *adjective*

cast fos·sil (kast fos' əl) *noun* A fossil formed when a mold left by a living thing fills with minerals that harden. *Most dinosaur "bones" that we study are really cast fossils of dinosaur bones.*

castings ▶ chemical formula

cast·ings (kast' iŋz) *noun, plural*
The wastes left in the soil by earthworms that have fed on decaying plants and animals. *We learn about earthworms by studying their castings.*

cause (koz) *noun* The reason something happens. *Scientists may study the causes of pollution.* **cause** *verb*

cell (sel) *noun*
The smallest unit of life. *Plants and animals are made up of cells that do many different jobs.*
cellular *adjective*

cell mem·brane (sel mem' brān) *noun*
The outer covering of a cell. *The cell membrane protects and gives shape to the cell.*

cell wall (sel wôl) *noun* The stiff covering outside the cell membrane of all plant cells. *Rigid cell walls allow a stem to stand up.*

Cel·si·us (sel' sē əs) *noun* The metric temperature scale; often abbreviated C. *The boiling point of water is 100 degrees Celsius.*

Ce·no·zo·ic era (sē nə zō' ik îr' ə) *noun* The current period of Earth's history. *The Cenozoic era began about 65 million years ago, following the age of dinosaurs.*

cen·ti·me·ter (sen' ti mē tər) *noun*
A metric unit of length equal to 1/100 of a meter. *A centimeter is about the width of a fingernail, a little shorter than $^1/_2$ inch.*

cer·e·bel·lum (ser ə bel' əm) *noun*
The part of the brain that controls body movement and balance. *The cerebellum sends messages to the muscles so they will work together.*

cer·e·brum (ser' ə brəm, sə rē' brəm) *noun* The part of the brain that controls thought, feelings, and memory. *The cerebrum is the largest part of the brain, filling much of the skull.*

charge (chärj) *noun* The amount of electricity in an object. *Objects may have a negative charge or a positive charge.*

chem·i·cal change (kem' i kəl chānj) *noun* A change that forms a different kind of matter. *Burning wood is an example of a chemical change.*

chemical energy (kem' i kəl en' ər jē) *noun* Stored energy that is given off or taken in during a chemical change. *When a match is lit, its chemical energy is given off as light and heat.*

chemical for·mu·la (kem' i kəl fôr' myə lə) *noun* A group of symbols and numbers that show the chemical makeup of a compound. *The chemical formula for water is H_2O.*

a	cat	e	net	îr	gear	u	cup	u̇	look, pull	th	this	ə	alive,
ā	day, lake	ē	seed	o	hot	ū	fuse	oi	soil	hw	wheel		comet,
ä	father	i	fit	ō	cold	ûr	fur, bird	ou	out	zh	measure		acid, atom,
âr	dare	ī	pine	ô	paw	ü	tool, rule	th	thin	ŋ	wing		focus

chemical prop·er·ty (kem' i kəl prop' ər tē) *noun* The way a chemical substance affects other substances. *A chemical property of oxygen is that it can make iron rust.*

chemical re·ac·tion (kem' i kəl rē ak' shən) *noun* A process that changes one or more substances into other substances. *The chemical reaction of hydrogen and oxygen can create water.*

chemical sym·bol (kem' i kəl sim' bəl) *noun* The letter code for an element. *The chemical symbol for silver is Ag.*

chem·is·try (kem' is trē) *noun* The scientific study of chemical elements and compounds. *Chemistry is the study of what substances are made of and how they change.*

chi·tin (kī' tən) *noun* The hard, horny material that forms the covering of some animals. *Lobster shells are made of chitin.*

chlo·rine (klôr' ēn) *noun* A greenish yellow, poisonous gas. *Chlorine can be used to purify water.*

chlo·ro·phyll (klôr' ə fil) *noun* The green coloring in plants. *Plants use chlorophyll to capture sunlight to make food.*

chlo·ro·plast (klôr' ə plast) *noun* The part of a plant cell that contains chlorophyll. *Chloroplasts are where photosynthesis takes place.*

chro·mo·some (krō' mə sōm) *noun* The part of the cell nucleus that carries the genes that give living things their characteristics. *Chromosomes pass on the traits that people get from their parents.*

chrys·a·lis (kris' ə ləs) *noun* A butterfly pupa, between the larva and adult stages. *A chrysalis is covered by a hard outer case.*

cir·cuit (sûr' kit) *noun* The complete path of an electric current. *A circuit must be closed for electric current to flow.*

circuit break·er (sûr' kit brā' kər) *noun* A switch that stops the flow of electricity in a circuit when the current is too high. *Without a circuit breaker, a circuit could become overloaded and start a fire.*

cir·cu·la·tion (sûr kyə lā' shən) *noun* The movement of blood through arteries and veins. *Circulation of blood carries oxygen and food to all parts of the body, and removes carbon dioxide and waste.* **circulatory** *adjective*

cir·cu·la·to·ry sys·tem (sûr' kyə lə tôr ə sis' təm) *noun* The organ system that moves blood through the body. *The circulatory system includes the heart and blood vessels.*

cir·rus cloud (sîr' əs kloud) *noun* A high-altitude cloud with a wispy, featherlike shape. *Cirrus clouds are made mostly of ice crystals.*

class (klas) *noun* A scientific grouping of animals within a phylum. *Mammals are the class of animals that have fur or hair and bear live young.*

clas·si·fi·ca·tion (klas i fi kā' shən) *noun* The sorting of things into groups based on ways they are alike. *The classification system for living things has five main groups, called kingdoms.*

clas·si·fy (klas' i fī) *verb* To put things into groups according to their properties. *You can classify objects as solid, liquid, or gas.*

clay (klā) *noun* A type of soil made up of very small grains. *Clay can hold water without leaking and becomes hard when baked.*

cli·mate (klī' mət) *noun* The usual weather conditions of a place over a long period of time. *The rain forest has a tropical climate.*

clone (klōn) *verb* To grow an identical plant or animal from a cell of a parent. *Scientists cloned a sheep in 1997 and named the new sheep Dolly.* **clone** *noun*

closed cir·cuit (klôzd sûr' kit) *noun* An electrical path that has no breaks or gaps. *Electric current can flow only through a closed circuit.*

cloud (kloud) *noun* A white or gray mass of tiny water droplets or ice crystals that floats in the air above Earth. *Clouds form when the air cools and water vapor in it turns back to liquid.*

coal (kōl) *noun* A soft, black mineral often burned as fuel. *Coal formed from plants that lived hundreds of millions of years ago.*

coarse (kôrs) *adjective* Having a rough texture or made of large particles. *Gravel is more coarse than sand.*

coast·al for·est (kōst' əl for' əst) *noun* A thick forest with tall trees that grows in temperate climates near the ocean. *The coastal forests of the Pacific Northwest contain giant redwood trees.*

coch·le·a (kok' lē ə) *noun* A coiled tube in the inner ear, lined with tiny hairs. *Nerve endings in the cochlea change sound wave vibrations into messages for the brain.*

co·coon (kə kün') *noun* A silky case made by some organisms for protection. *Some spiders wrap their eggs in cocoons.*

co·he·sion (kō hē' zhən) *noun* The attraction that holds the particles of a substance together. *Cohesion is stronger in solids than in gases.* **cohere** *verb* **cohesive** *adjective*

a	cat	e	net	îr	gear	u	cup	u̇	look, pull	*th*	this	ə	alive,
ā	day, lake	ē	seed	o	hot	ū	fuse	oi	soil	hw	wheel		comet,
ä	father	i	fit	ō	cold	ûr	fur, bird	ou	out	zh	measure		acid, atom,
âr	dare	ī	pine	ô	paw	ü	tool, rule	th	thin	ŋ	wing		focus

cold-blood·ed (kōld′ blud əd) *adjective* Having a body temperature that is warmed or cooled by the surrounding air or water. *Fish and snakes are cold-blooded animals.*

cold front (kold frunt) *noun* A weather boundary formed when a mass of cold air moves under a mass of warm air. *Cold fronts often bring stormy weather.*

col·o·ny (kol′ ə nē) *noun* A group of animals or plants living and growing together. *The colony of ants built a huge anthill.*

com·et (kom′ ət) *noun* A frozen mass of ice and dust that orbits the Sun. *As a comet nears the Sun, it looks like a bright ball with a long, glowing tail.*

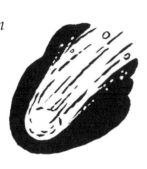

com·mu·ni·cate (kə mū′ ni kāt) *verb* To share information. *Be sure to communicate the results of your experiment.*

com·mu·ni·ty (kə mū′ ni tē) *noun* All the populations of organisms that live in the same area. *Plants and animals in a community depend on one another for survival.*

com·pare (kəm pâr′) *verb* To identify how things are alike. *A graph can help us compare results.* **comparison** *noun*

com·pass (kum′ pəs) *noun* An instrument that points to magnetic north. *The magnetic needle on a compass is attracted to Earth's north magnetic pole.*

com·pe·ti·tion (kom pə tish′ ən) *noun* The struggle among living things for food, water, or other needs. *The competition for water increases during a drought.*

com·plex ma·chine (kom pleks′ mə shēn′) *noun* A machine for doing work made up of two or more simple machines and powered by something other than human or animal effort. *A motor is a complex machine.*

com·pound (kom′ pound) *noun* Two or more elements or materials bonded together. *Water is a compound of hydrogen and oxygen.*

compound ma·chine (kom′ pound mə shēn′) *noun* Two or more simple machines put together. *A bicycle is a compound machine.*

com·press (kəm pres′) *verb* To make something smaller by squeezing or pressing it. *You can compress a foam ball by holding it tightly in your fist.*

com·pres·sion (kəm presh′ ən) *noun* The part of a sound wave in which air is pushed together. *A musical instrument that makes low sounds has compressions that are far apart.*

con·cave (kon kāv′) *adjective* Curved inward like the inside of a bowl. *A concave lens is thin in the middle and wide at the edges.*

con·cus·sion (kən kush′ ən) *noun* An injury to the brain caused by a blow to the head. *My friend slipped and fell on the ice and got a concussion.*

con·den·sa·tion (kon den sā′ shən) *noun* The change from gas to liquid as a result of cooling. *Condensation causes dew to form on leaves and grass as the air temperature goes down.*

con·dense (kən dens′) *verb* To change from a gas to a liquid state. *When water vapor hits a cool surface, it condenses into water droplets.*

con·duc·tion (kən duk′ shən) *noun* The transfer of energy through a solid material. *Heat conduction happens when particles of matter strike other particles speeding up their motion.* **conduct** *verb*

con·duc·tor (kən duk′ tər) *noun* A material that heat or electrical energy can pass through easily. *Copper is a good conductor of electricity.*

cone (kōn) *noun* 1. The part of a conifer that produces seeds. *The woody scales of the pine cone protect the seeds inside.* 2. An object or shape with a round base and pointed top. *Volcanoes are often shaped like cones.*

con·glom·er·ate (kən glom′ ə rət) *noun* A rock formed from pebble-sized rock pieces cemented together. *Conglomerate is a type of sedimentary rock.*

con·i·fer (kon′ ə fər) *noun* An evergreen tree that produces seeds inside cones. *Pine, fir, and spruce trees are all conifers.*

co·nif·er·ous for·est (kə nif′ ə rəs fôr′ əst) *noun* A large growth of evergreen trees. *Coniferous forests once covered much of New England.*

con·ser·va·tion (kon sûr vā′ shən) *noun* The protection of natural resources such as rivers and forests. *Many environmental groups work for the conservation of wilderness areas.*

con·stel·la·tion (kon stə lā′ shən) *noun* A group of stars that form a pattern in the sky as seen from Earth. *Many constellations are named after animals or figures from myths.*

con·sum·er (kən sü′ mər) *noun* An organism that survives by eating other organisms. *Almost all producers are plants, but consumers are usually animals.*

con·text (kon′ tekst) *noun* The background conditions in which something happens. *Biologists who study animals in the wild watch the animals in context.*

con·ti·nent (kon′ ti nent) *noun* One of the seven large land masses on Earth. *The continents are North America, South America, Asia, Europe, Africa, Australia, and Antarctica.*

a	cat	e	net	îr	gear	u	cup	u̇	look, pull	th	this	ə	alive,
ā	day, lake	ē	seed	o	hot	ū	fuse	oi	soil	hw	wheel		comet,
ä	father	i	fit	ō	cold	ûr	fur, bird	ou	out	zh	measure		acid, atom,
âr	dare	ī	pine	ô	paw	ü	tool, rule	th	thin	ŋ	wing		focus

con·ti·nen·tal drift (kon ti nen′ təl drift) *noun* An early theory stating that Earth's crust is made of huge plates that moved to their present positions. *The theory of continental drift does not explain how the continents were able to move.*

continental shelf (kon ti nen′ təl shelf) *noun* The land around the edge of a continent that is under water. *The continental shelf slopes away from the coastline for 10 to 1200 kilometers.*

continental slope (kon ti nen′ təl slōp) *noun* The steep drop-off in the seafloor beyond the continental shelf. *The ocean floor begins to flatten at the base of the continental slope.*

con·tour plow·ing (kon′ tür plou′ iŋ) *noun* Farming a piece of land so the rows follow the shape of the land. *Contour plowing reduces soil erosion.*

con·tract (kən trakt′) *verb* To get smaller. *Matter often contracts when it is cooled.*

con·trast (kən trast′) *verb* To identify how things are different. *You would contrast stars and planets by explaining how they are not alike.*

con·trol (kən trōl′) *noun* The part of an experiment that stays the same while other parts change. *The control group of pea plants received water every day.* **control** *verb*

con·vec·tion (kən vek′ shən) *noun* The transfer of heat by currents through a liquid or gas. *A hot air balloon is an example of convection at work.*

con·vex (kon′ veks) *adjective* Curved outward like the outside of a ball. *Many magnifiers are convex lenses.*

cor·al reef (kôr′ əl rēf) *noun* Underwater formations made of the hard skeletons of tiny sea animals. *We got a chance to snorkel at a coral reef.*

core (kôr) *noun* The center of the planet. *The core of Earth is superhot.*

cor·ne·a (kôr′ nē ə) *noun* The clear outer layer of the eyeball. *The cornea covers the iris and pupil.*

co·ro·na (kə rō′ nə) *noun* The outer atmosphere of a star. *The Sun's corona can be seen during a total solar eclipse.*

cot·y·le·don (kot ə lē′ dən) *noun* The part of a seed that provides food for the new plant; also called the seed leaf. *Some cotyledons become the first leaves of the new plant.*

cra·ter (krā′ tər) *noun* A bowl-shaped hollow or depression. *Meteorites created huge craters on the Moon's surface.*

cres·cent (kres′ ənt) *noun* The first or last phase of the Moon's cycle. *A crescent Moon looks like the end of a fingernail.*

crust (krust) *noun* The hard outer covering of Earth. *Earthquakes happen when parts of Earth's crust slip or move.*

crys·tal (kris′ təl) *noun* A solid substance with a regular pattern of flat surfaces. *When ocean water evaporates, salt crystals are left.* **crystalline** *adjective*

cu·bic me·ter (kū' bik mē' tər) *noun*
A metric measure of volume. *A cubic meter is 1 meter wide, 1 meter high, and 1 meter deep.*

cul·ture (kul' chər) *noun* The laboratory growth of microbes in a specially prepared food. *Scientists use cultures to make vaccines for certain diseases.*

cu·mu·lus cloud (kū' myə ləs kloud) *noun* A puffy, middle-altitude cloud with a flat base and rounded top. *Small cumulus clouds are usually signs of fair weather.*

cur·rent (kûr' ənt) *noun* 1. A flow of electricity through a circuit. *Use the switch to control the current.* 2. A stream of water flowing in a body of water. *Surface currents in the ocean are caused by winds.*

current elec·tric·i·ty (kûr' ənt i lek tris' i tē) *noun* Electricity that flows through a circuit. *The two kinds of electricity are current electricity and static electricity.*

cy·cle (sī' kəl) *noun* A process that repeats itself in the same order over time. *The water cycle is the ongoing movement of water on, below, and above Earth's surface.*

cy·to·plasm (sī' tō plaz əm) *noun* A clear, jellylike material that fills plant and animal cells. *The nucleus and other parts of a cell are in the cytoplasm.*

dark mat·ter (därk mat' ər) *noun* Invisible substance thought to make up most of the mass of the universe. *Scientists estimate that 90 percent of the universe may be dark matter.*

da·ta (dat' ə, dā' tə) *noun, plural* Information or facts. *Meteorologists collect data about weather.* **datum** *singular*

dead (ded) *adjective* No longer alive. *The remains of dead plants and animals are broken down and returned to the soil.*

de·cid·u·ous for·est (də sij' ü əs fôr' əst) *noun* A forest of trees that lose and regrow their leaves every year. *A deciduous forest might contain maple, oak, and birch trees.*

de·com·pos·er (dē kəm pō' zər) *noun* A living thing that breaks down dead plant and animal materials. *Fungi and bacteria are examples of decomposers.*

a	cat	e	net	îr	gear	u	cup	u̇	look, pull	*th*	**this**	ə	alive,
ā	day, lake	ē	seed	o	hot	ū	fuse	oi	soil	hw	**wheel**		comet,
ä	father	i	fit	ō	cold	ûr	fur, bird	ou	**out**	zh	measure		acid, atom,
âr	dare	ī	pine	ô	paw	ü	tool, rule	th	**thin**	ŋ	wing		focus

deep o·cean cur·rent (dēp ō' shən kûr' ənt) *noun* A stream of water that flows below the ocean surface. *Cold water sinks beneath warm water creating deep ocean currents.*

de·gree (də grē') *noun* A unit for measuring temperature. *Water freezes at 32 degrees Fahrenheit (0 degrees Celsius).*

del·ta (del' tə) *noun* The flat area of land where a river drops soil and sand as it flows into the ocean. *A river delta is often shaped like a triangle.*

den·si·ty (den' si tē) *noun* The amount of matter in a given space. *The density of water is greater than the density of oil.* **dense** *adjective*

de·pen·dence (di pen' dəns) *noun* A need for something for support or survival. *Plants and animals that share a habitat have a dependence on one another.* **depend** *verb* **dependent** *adjective*

de·pres·sant (di pres' ənt) *noun* A substance that lessens pain, causes sleep, or lowers anxiety. *Depressants slow down the body's activities.*

der·mis (dûr' mis) *noun* The inner layer of the skin. *Nerve endings and blood vessels are in the dermis.*

des·ert (dez' ərt) *noun* A region where there is very little rain. *The cactus is a plant that can grow in the desert.*

de·vel·op·ment (di vel' əp mənt) *noun* The way a living thing changes during its life. *Most animals go through a stage of growth and development after birth.* **develop** *verb*

di·a·phragm (dī' ə fram) *noun* The wall of muscle between the chest and abdomen. *Movement of the diaphragm controls breathing.*

di·cot seed (dī' kot sēd) *noun* A plant seed that develops two cotyledons, or seed leaves. *Many flowering plants produce dicot seeds.*

di·ges·tion (dī jes' chən) *noun* The breaking down of food into tiny pieces that can be used by the body. *Digestion begins as soon as our teeth begin to chew food.* **digest** *verb*

di·ges·tive sys·tem (dī jes' tiv sis' təm) *noun* The organ system that breaks down food so it can enter the bloodstream. *The digestive system includes the mouth, stomach, intestines, liver, and other organs.*

di·lute (dī lüt') *verb* To make a mixture weak by adding water. *The ice cubes will dilute the juice when they melt.*

di·no·saurs (dī' nə sôrz) *noun, plural* A group of animals that lived on Earth millions of years ago. *Some of the largest dinosaurs were plant eaters.*

dir·ect cur·rent (də rekt' kûr' ənt) *noun* Electric current that flows in one direction through a circuit. *Batteries produce direct current.*

15

dis·ease (di zēz') *noun* Sickness or ill health. *Measles was once a common childhood disease.*

dis·solve (di zôlv') *verb* To form a solution when added to a liquid. *Both sugar and salt dissolve in water.*

dis·tance (dis' təns) *noun* How far apart two points are. *The distance between Chicago and Dallas is about 1,480 kilometers (920 miles).*

di·ver·si·ty (di vûr' si tē) *noun* Difference or variety. *A rain forest has a great diversity of animals.* **diverse** *adjective*

dome moun·tains (dōm moun' tənz) *noun, plural* Mountains formed by forces pushing Earth's crust upward into bulges. *The Black Hills of South Dakota and Wyoming are dome mountains.*

dor·mant (dôr' mənt) *adjective* Alive but not active or growing. *The seeds were dormant until the rains came.*

drone bee (drōn bē) *noun* A male bee that has no stinger and does no work. *Only drone bees mate with the queen.*

drought (drout) *noun* A long period of unusually dry weather. *Many crops died during the drought.*

drug (drug) *noun* A substance that treats illness or that is taken for another effect. *Cough medicine is a kind of drug you might take when you have a cold.*

drum·lin (drum' lin) *noun* An oval hill of sand and gravel formed by a glacier. *The drumlins in Wisconsin were created during the last Ice Age.*

dry cell (drī sel) *noun* An energy source with chemicals stored in paste form so they will not spill. *The dry cells used in toys and flashlights are often called batteries.*

duc·tile (duk' təl) *adjective* Easy to hammer thin or to form into a new shape. *Copper wire bends easily because copper is a ductile metal.*

dune (dün) *noun* A ridge of sand piled up by the wind. *Behind the beach was a long row of dunes.*

Dust Bowl (dust' bōl) *noun* The large region of the American Great Plains that suffered a long drought and dust storms during the 1930s. *The wind blew away millions of tons of topsoil in the Dust Bowl.*

a	cat	e	net	îr	gear	u	cup	ú	look, pull	*th*	**th**is	ə	alive,
ā	day, lake	ē	seed	o	hot	ū	fuse	oi	soil	hw	**wh**eel		comet,
ä	father	i	fit	ō	cold	ûr	**fur, bird**	ou	**out**	zh	mea**s**ure		acid, atom,
âr	dare	ī	pine	ô	paw	ü	tool, rule	th	**th**in	ŋ	wi**ng**		focus

ear bones (îr bōnz) *noun, plural*
The three bones in the middle ear, just behind the eardrum. *The ear bones, called the hammer, anvil, and stirrup, are the smallest bones in the body.*

ear·drum (îr′ drum) *noun* The thin layer of tissue stretched between the outer and middle ear. *Sound waves make the eardrum vibrate.*

Earth (ûrth) *noun*
The planet on which we live. *Earth is the third planet from the Sun in the solar system.*

earth·quake
(ûrth′ kwāk) *noun* A shaking movement of the ground caused by a sudden shift of Earth's crust. *Many earthquakes have occurred in California, Japan, and Mexico.*

e·chi·no·derm (i kī′ nə dûrm) *noun*
A spiny-skinned animal that has no backbone. *Sea urchins and starfish are echinoderms.*

ech·o (ek′ ō) *noun* A sound that is heard again after it bounces back from a surface. *If you shout in an empty hallway, you will hear the echo of your voice.*
echo *verb*

ech·o·lo·ca·tion (ek ō lō kā′ shən) *noun*
Finding objects by listening to their echoes. *Bats and dolphins use echolocation to find food.*

e·clipse (ē klips′) *noun* The partial or complete hiding of one body in space by another. *In a solar eclipse, the Sun is hidden from view by the Moon.*

e·col·o·gist (ē kol′ ə jist) *noun*
A scientist who studies ecology, or the relationships among living and nonliving things. *Ecologists warn against destroying natural habitats.*

e·col·o·gy (ē kol′ ə jē) *noun* The study of how living and nonliving things interact in the environment. *Ecology is concerned with the ways populations of organisms adapt and survive.*

ec·o·sys·tem (ek′ ō sis təm) *noun*
A community of living and nonliving things in an environment and all their interactions. *The organisms in an ecosystem depend on one another to survive.*

ef·fect (ə fekt′) *noun* A result or outcome. *Scientists often study causes and effects.*

ef·fort force (ef′ ərt fôrs) *noun*
The amount of force needed to move something. *Simple machines reduce the effort force required to do work.*

egg (eg) *noun* An oval or round body covered by a shell by which some animals reproduce. *Birds and crocodiles lay eggs.*

e·lec·tri·cal en·er·gy
(i lek′ tri kəl en′ ər jē) *noun* The energy of moving electric charges. *Electrical energy is used to run computers, TVs, and many other things we use every day.*

e·lec·tric cell (i lek′ trik sel) *noun* A device in which a chemical reaction produces an electric current. *Dry cells and car batteries are electric cells.*

electric charge (i lek′ trik chärj) *noun* 1. An amount of electricity. *The electric charge in lightning is greater than the electric charge in a car battery.* 2. A single positive or negative charge. *The continuous movement of electric charges creates electric current.*

electric cir·cuit (i lek′ trik sûr′ kit) *noun* The path along which an electric current flows. *Current can flow only if the electric circuit is closed, or unbroken.*

electric cur·rent (i lek′ trik kûr′ ənt) *noun* The flow of electric charges through a substance in a closed circuit. *An electric current flows from the power source through wires.*

electric dis·charge (i lek′ trik dis′ chärj) *noun* The sudden movement of an electrical charge from the point of buildup to another point. *Lightning is an electric discharge in the clouds.*

electric field (i lek′ trik fēld) *noun* The area around a charged object where the effects of electricity can be observed.

e·lec·tric·i·ty (i lek tris′ ə tē) *noun* A form of energy produced by the movement of electric charges. *Most household appliances are powered by electricity.*

e·lec·trode (i lek′ trōd) *noun* A conductor that carries electric current into or out of an electrical device. *Batteries have two electrodes, one positive and one negative.*

e·lec·tro·mag·net (i lek trō mag′ nət) *noun* A temporary magnet created when current flows through wires coiled around a piece of iron. *An electromagnet loses its magnetism when the current is turned off.*

e·lec·tron (i lek′ tron) *noun* A tiny particle that orbits as part of an atom and that moves around outside the nucleus. *Electrons carry a negative electric charge.*

e·le·ment (el′ ə mənt) *noun* A substance that is made up of only one type of matter. *Elements, such as oxygen, lead, and gold, cannot be broken down into other substances.*

el·lipse (i lips′) *noun* An oval shape. *The shape of Earth's orbit around the Sun is an ellipse.* **elliptical** *adjective*

a	cat	e	net	îr	gear	u	cup	u̇	look, pull	th	this	ə	alive,
ā	day, lake	ē	seed	o	hot	ū	fuse	oi	soil	hw	wheel		comet,
ä	father	i	fit	ō	cold	ûr	fur, bird	ou	out	zh	measure		acid, atom,
âr	dare	ī	pine	ô	paw	ü	tool, rule	th	thin	ŋ	wing		focus

em·bry·o (em′ brē o) *noun* A very young organism just beginning to grow. *The embryo of a plant is contained in the seed and grows when the seed sprouts.*

en·dan·gered (en dān′ jûrd) *adjective* In danger of dying off, or becoming extinct. *The manatee is an endangered species.*

en·do·skel·e·ton (en dō skel′ i tən) *noun* A supporting structure of bones inside the body of an animal. *Mammals, birds, fish, and reptiles have endoskeletons.*

en·er·gy (en′ ər jē) *noun* The ability to do work. *Animals get energy from the food they eat.*

energy pyr·a·mid (en′ ər jē pîr′ ə mid) *noun* A diagram that shows how energy is used in an ecosystem. *Producers are at the bottom of every energy pyramid.*

energy trans·for·ma·tion (en′ ər jē trans fôr mā′ shən) *noun* The change of energy from one form to another; also called energy transfer. *An example of energy transformation is light energy from the Sun changing to electrical energy in a solar calculator.*

en·vi·ron·ment (en vī′ rən mənt) *noun* Everything around an organism that affects it. *Climate, land, air, water, and food sources are all part of our environment.* **environmental** *adjective*

en·zyme (en′ zīm) *noun* A protein in the body that starts or speeds up a chemical reaction. *Enzymes in the stomach help digest food.*

ep·i·cen·ter (ep′ i sen tər) *noun* The point on Earth's surface directly above where an earthquake starts. *Earthquake waves are strongest close to the epicenter.*

ep·i·der·mis (ep i dûr′ mis) *noun* The outermost protective layer of cells or tissue on an organism. *The epidermis of an animal is its skin.*

e·qua·tor (ē kwā′ tər) *noun* The imaginary line around the middle of a planet, halfway between the poles. *The climate is very hot near Earth's equator.*

e·qui·nox (ē′ kwə noks) *noun* One of two times during Earth's revolution when the Sun is directly overhead in the sky at the equator. *On the fall and spring equinoxes, day and night are equal in length.*

er·a (îr′ ə) *noun* A long period of time. *Earth's history is divided into three main eras: Paleozoic, Mesozoic, and Cenozoic.*

e·ro·sion (i rō′ zhən) *noun* The gradual wearing away of soil and rock. *Water and wind are the main forces of erosion.* **erode** *verb*

er·rat·ic (i rat′ ik) *noun* A boulder left behind by a glacier. *The huge rock in the middle of the field is an erratic.*

e·soph·a·gus (i sof′ ə gəs) *noun* The tube through which food passes from the mouth to the stomach. *The esophagus is in the throat, behind the windpipe.*

es·ti·mate (es' tə māt) *verb* To make a rough guess of size or value based on some evidence. *The scientist estimated the fossil was 10,000 years old.* **estimate** *noun*

es·tu·ary (es' chü er ē) *noun* The wide mouth of a river where it flows into the ocean. *Tides make the water in an estuary a mix of salt water and fresh water.*

e·vap·o·rate (i vap' ər āt) *verb* To change from a liquid to a gas. *Liquids evaporate when they are heated to their boiling points.*

e·vap·o·ra·tion (i vap ər ā' shən) *noun* The change of a liquid to a gas. *Evaporation made the puddle disappear in a day.*

e·vo·lu·tion (ev ə lü' shən) *noun* The change in species over long periods of time. *Fossils show the evolution of the horse from a small, dog-sized animal to what we see today.* **evolve** *verb*

ex·cre·to·ry sys·tem (ek' skri tôr ē sis' təm) *noun* The organ system that removes liquid wastes from the body. *The kidneys and bladder are part of the excretory system.* **excrete** *verb*

ex·hale (eks hāl') *verb* To breathe out. *When you exhale, carbon dioxide is squeezed out of your lungs.*

ex·o·skel·e·ton (ek sō skel' i tən) *noun* A hard, stiff covering that protects the body of an invertebrate. *Lobsters have an exoskeleton.*

ex·pand (ek spand') *verb* To become larger. *When you breathe in, your lungs expand.*

ex·per·i·ment (ek sper' ə mənt) *noun* A set of trials that tests a hypothesis. *The students did an experiment to find out what crickets eat.* **experiment** *verb*

ex·pi·ra·tion date (ek spə rā' shən dāt) *noun* The date printed on a label that tells when to throw out the product and not use it. *A medicine may not work after its expiration date.*

ex·ter·nal (ek stûr' nəl) *adjective* On the outside of the body. *A clam's shell is an external covering.*

ex·tinct (ek stiŋkt') *adjective* No longer alive on Earth. *Dinosaurs became extinct about 65 million years ago.*

eye·ball (ī' bôl) *noun* The sphere-shaped sense organ for sight. *The cornea, iris, pupil, lens, and retina are parts of the eyeball.*

a	cat	e	net	îr	gear	u	cup	u̇	look, pull	*th*	**this**	ə	alive,
ā	day, lake	ē	seed	o	hot	ū	fuse	oi	soil	hw	**wheel**		comet,
ä	father	i	fit	ō	cold	ûr	fur, bird	ou	**out**	zh	measure		acid, atom,
âr	dare	ī	pine	ô	paw	ü	tool, rule	th	thin	ŋ	wing		focus

eye·piece (ī′ pēs) *noun* The part of a microscope or telescope that you look through. *A lens is usually part of an eyepiece.*

fac·tor (fak′ tər) *noun* A cause or event that brings about a result. *Wind and temperature are two important weather factors.*

Fah·ren·heit (fâr′ ən hīt) *adjective* The customary temperature scale used in the United States; often abbreviated F. *Normal body temperature for humans is 98.6 degrees Fahrenheit.*

false legs (fôls legz) *noun, plural* The back pairs of legs of a larva, used for crawling. *A caterpillar's false legs disappear during metamorphosis, and only the six true legs remain.*

fam·i·ly (fam′ ə lē) *noun* A smaller group of plants and animals within a class. *Cats, lynxes, and lions belong to the feline family.*

fat (fat) *noun* Oily substance found in animal bodies and plant seeds. *Meats, milk, and nuts are high in fat.*

fault (fôlt) *noun* A crack or break in Earth's outer crust along which movement occurs. *Earthquakes often happen along faults.*

fault-block moun·tains (fôlt′ blok moun′ tənz) *noun, plural* Mountains formed when huge blocks of rock are tilted up along a crack in Earth's crust. *The Sierra Nevada in California are fault-block mountains.*

fea·ture (fē′ chər) *noun* A general body structure. *The eyes, ears, nose, and mouth are features of the face.*

fe·mur (fē′ mər) *noun* The thigh bone. *The femur is the longest bone in the human body.*

fer·tile (fûr′ təl) *noun* Able to produce fruit, seeds, or young. *A blueberry bush that makes berries is fertile.*

fer·til·i·za·tion (fûr təl lə zā′ shən) *noun* The joining of a male and female plant cell to form a seed. *Fertilization takes place in the ovary of a flower.* **fertilize** *verb*

fer·ti·li·zer (fûr′ tə lī zər) *noun* Substance added to soil to help plants grow. *Some fertilizers can pollute streams and lakes.*

fi·ber (fī′ bər) *noun* The part of foods that cannot be digested. *The fiber in fruits and vegetables helps move food through the digestive system.*

fi·brous root (fī′ brəs rüt) *noun* A hairy, branching plant root. *The two kinds of roots are fibrous roots and taproots.*

fil·a·ment
(fil' ə mənt) *noun* A very fine wire that glows when electric current goes through it. *Most light bulb filaments are made of tungsten.*

fil·ter
(fil' tər) *noun* A material that cleans liquids or gases that pass through it. *Sand can act as a filter for water.*
filter *verb*

fil·tra·tion
(fil trā' shən) *noun* Using a filter to remove particles or pollutants. *The filtration system makes our water safe to drink.*

fish
(fish) *noun* A cold-blooded animal with a backbone that lives in water. *Fish breathe through gills, not lungs, and are covered with scales.*

fixed pul·ley
(fikst púl' ē) *noun* A simple machine made of a wheel held in place with a rope going over it. *Fixed pulleys are used to raise and lower flags.*

flash flood
(flash flud) *noun* A rapidly rising rush of water over normally dry land. *The rainstorm produced flash floods in the canyon.*

flood
(flud) *noun* A large amount of water that covers land that is usually dry. *The river overflows its banks in a flood every spring.* **flood** *verb*

flow·er·ing plant
(flou' ər iŋ plant) *noun* A plant that produces seeds after making flowers. *Flowering plants often have brightly colored petals.*

fluo·res·cent
(flü res' ənt) *adjective* Giving off light when electric current makes gas inside a tube glow. *The lights in the classroom ceiling are fluorescent lights.*

fo·cus
(fō' kəs) *noun* A point where rays of light come together after going through a lens. *The focus points in the eye are on the retina.*

fold·ed moun·tains
(fōl' dəd moun' tənz) *noun, plural* Mountains formed when layers of rock are pushed upward from the sides. *The Appalachians and the Alps are folded mountains.*

food chain
(füd chān) *noun* A way to describe which animals eat other organisms in an ecosystem. *A cat eating a bird that ate sunflower seeds is an example of a food chain.*

food web
(füd web) *noun* A model that shows how food chains connect and overlap. *All the food chains in a community make up a food web.*

force
(fôrs) *noun* A push or a pull. *Work is done when a force moves an object a distance.*

a	cat	e	net	îr	gear	u	cup	ú	look, pull	*th*	this	ə	alive,
ā	day, lake	ē	seed	o	hot	ū	fuse	oi	soil	hw	wheel		comet,
ä	father	i	fit	ō	cold	ûr	fur, bird	ou	out	zh	measure		acid, atom,
âr	dare	ī	pine	ô	paw	ü	tool, rule	th	thin	ŋ	wing		focus

fore·cast (fôr′ kast) *verb* To predict what will happen. *Scientists study temperature, wind, air pressure, and clouds to forecast the weather.* **forecast** *noun*

for·est (fôr′ əst) *noun* A thick growth of trees covering a large area of land. *Different kinds of forests grow in different climates.*

fos·sil (fos′ əl) *noun* The remains of an organism that lived long ago. *Studying fossils helps us learn about the past.*

fos·sil fuel (fos′ əl fū′ əl) *noun* Fuel formed from the remains of plants and animals that lived long ago. *Coal and petroleum are fossil fuels.*

frac·ture (frak′ chər) *verb* To break or crack, especially a bone or rock. *She fell and fractured her wrist bone.* **fracture** *noun*

frame of ref·er·ence (frām əv ref′ ər əns) *noun* A description of where an object is, compared with objects around it. *The frame of reference for Earth in the solar system is the orbits of other planets.*

freeze (frēz) *verb* To change from liquid to solid. *When water freezes, it turns to ice.*

freez·ing point (frēz′ iŋ point) *noun* The temperature at which a liquid becomes a solid. *The freezing point of water is 0° C (32° F).*

fresh wa·ter (fresh wô′ tər) *noun* Water that has very little salt in it. *Rain is fresh water.*

fric·tion (frik′ shən) *noun* A force caused when two objects rub together, slowing their motion. *Smooth surfaces produce less friction than rough surfaces.*

front (frunt) *noun* The boundary where two air masses of different temperatures meet. *A cold front usually brings rain followed by cooler weather.*

fruit (früt) *noun* The part of a flowering plant that contains seeds. *Grapes, apples, and melons are fruits.*

fuel (fū′ əl) *noun* A substance burned to provide energy. *Car engines use gasoline as fuel.*

ful·crum (fŭl′ krəm) *noun* The fixed point on which a lever turns or pivots. *The closer the fulcrum is to the load, the less force is needed to lift the load.*

func·tion (fuŋk′ shən) 1. *noun* The work or job that a body part does. *The function of the heart is to pump blood.* 2. *verb* To work in a certain way. *The biceps muscle functions by getting shorter and pulling the lower arm.*

fun·gi (fun′ jī) *noun, plural* Plantlike organisms with no leaves, flowers, roots, or green color. *Fungi, such as mushrooms and toadstools, make up one of the five kingdoms of living things.* **fungus** *singular*

fun·nel (fun′ əl) *noun* An open cone that narrows to a tube. *Funnels are used for pouring liquids into containers.*

fuse (fūz) *noun* A device that melts to open a circuit and stop too much electric current from flowing through wires. *The lights went out when the fuse blew.*

gal·ax·y (gal' ək sē) *noun* A very large group of stars in space. *Our solar system is in the Milky Way galaxy.*

gas (gas) *noun* A state of matter that has no set shape and no set volume. *The air around us is a mixture of many gases.* **gaseous** *adjective*

gas gi·ant (gas jī' ənt) *noun* A huge planet made up mostly of gases. *Saturn and Jupiter are gas giants.*

gears (gērz) *noun, plural* Wheels with teeth that fit together, often part of a machine. *When one gear turns in one direction, the next gear turns in the opposite direction.*

gem·stone (jem' stōn) *noun* A mineral that can be cut and polished. *Emeralds, diamonds, and rubies are examples of gemstones.*

gene (jēn) *noun* A section of a chromosome that carries the information for a trait. *Genes are passed from parents to children.*

gen·er·al·ize (jen' ə rə līz) *verb* To make a general rule from a small number of examples. *After the experiment, the students generalized that peas were the fastest-growing plants.* **generalization** *noun* **general** *adjective*

gen·er·a·tion (jen ə rā' shən) *noun* The time period in which an organism is born, grows and develops, reproduces, and dies. *A generation equals one complete life cycle.*

gen·er·a·tor (jen' ər ā tər) *noun* A machine that produces electricity. *The generator turns energy from the waterfall into electrical energy.*

ge·nus (jē' nəs) *noun* A group of closely related plants or animals. *Foxes, wolves, and dogs belong to the same genus, called Canis.* **genera** *plural*

ge·ol·o·gist (jē ol' ə jist) *noun* Scientist who studies Earth and the materials it is made of. *Geologists study rocks to find out how Earth has changed over time.*

ge·o·ther·mal en·er·gy (jē ō thûr' məl en' ər jē) *noun* Heat energy from inside Earth. *Hot springs are warmed by geothermal energy.*

a	cat	e	net	îr	gear	u	cup	ు	look, pull	*th*	this	ə	alive,
ā	day, lake	ē	seed	o	hot	ū	fuse	oi	soil	hw	wheel		comet,
ä	father	i	fit	ō	cold	ûr	fur, bird	ou	out	zh	measure		acid, atom,
âr	dare	ī	pine	ô	paw	ü	tool, rule	th	thin	ŋ	wing		focus

germ (jûrm) *noun* A tiny organism that can cause disease. *Germs are too small to see with the naked eye.*

ger·mi·nate (jûr′ mə nāt) *verb* To sprout into a plant. *When seeds germinate, they grow shoots and roots.*

gills (gilz) *noun, plural* The breathing organ of fish and some other animals that live in water. *Gills look like slits behind the head of a fish.*

giz·zard (giz′ ərd) *noun* A part of the digestive system of some birds. *Tiny pebbles in the gizzard break down food that is hard to digest.*

gla·cial till (glā′ shəl til) *noun* A mixture of rocks left behind by a melting glacier. *Glacial tills show how far south the ice sheet came during the last Ice Age.*

gla·cier (glā′ shər) *noun* A huge mass of slowly moving ice. *Glaciers grow and shrink as climate changes.* **glacial** *adjective*

gland (gland) *noun* An organ that produces natural chemicals called hormones. *Glands control many body activities, including growth and digestion.*

glid·er (glī′ dər) *noun* An aircraft that has wings but no engine. *Gliders are carried by air currents.*

glo·bal warm·ing (glō′ bəl wôr′ miŋ) *noun* A theory that Earth's climate is getting warmer. *Some scientists think global warming is caused by pollution in the atmosphere trapping the Sun's heat energy.*

gold (gōld) *noun* A heavy, yellow metal that is an element. *Gold is often used for jewelry because it can be shaped easily and never rusts or tarnishes.*

grad·u·at·ed cyl·in·der (graj′ ü ā təd sil′ ən dər) *noun* A narrow, round container with straight sides marked with a scale to show volume. *Use the graduated cylinder that holds 100 milliliters.*

grain (grān) *noun* 1. The small, hard seed of some plants. *Flour is made from wheat grain.* 2. A very small particle. *Grains of sand, sugar, and salt are about the same size.*

gram (gram) *noun* The basic unit of mass in the metric system. *A new pencil has a mass of about 3 grams.*

gran·ite (gran′ it) *noun* A light-colored, large-grained igneous rock. *The steps in front of our school are made of granite.*

graph (graf) *noun* A diagram that shows data in lines or pictures. *Line graphs, bar graphs, and circle charts are three kinds of graphs.*

grass·land (gras′ land) *noun* An open, flat area covered by grasses but few trees. *The prairies of the U.S. Midwest are grasslands.*

grav·i·ty (grav′ i tē) *noun* The natural force that pulls objects toward each other. *Gravity causes objects to fall toward Earth's surface.* **gravitational** *adjective*

green·house ef·fect (grēn' hous ə fect') *noun* The warming of Earth's atmosphere as gases in the air trap the Sun's heat energy. *When we studied weather, we learned about the greenhouse effect.*

ground·ed (groun' dəd) *adjective* Connected in such a way that electricity flows into Earth. *An electrical appliance that is grounded is safe to use.*

ground·wa·ter (ground' wô tər) *noun* Fresh water stored in cracks and holes in underground rocks. *People drill wells to reach groundwater.*

hab·i·tat (hab' i tat) *noun* The place in nature where a plant or animal usually lives. *The habitat for lions is the African grasslands.*

hail (hāl) *noun* A kind of frozen precipitation. *Hail is small round pieces of ice that fall during a storm.* **hail** *verb*

hard·ness (härd' nəs) *noun* A material's resistance to being scratched. *The hardness of minerals is measured on Mohs' scale.* **hard** *adjective*

har·vest (här' vest) *verb* To collect the crop that has grown during a season. *They harvested the corn in August.* **harvest** *noun*

haz·ard·ous waste (haz' ər dəs wāst) *noun* Dangerous or poisonous materials that should be thrown away only with special handling. *We could not put the old cans of paint in the trash because they contain hazardous waste.*

head (hed) *noun* The front part of an insect's body. *An insect's mouthparts, eyes, and antennae are on its head.*

head·land (hed' lənd) *noun* Point of land stretching farthest into the ocean. *Lighthouses are often located on headlands.*

hear·ing (hîr' iŋ) *noun* The sense that receives sound. *The ears are the sense organ for hearing.*

hearing aid (hîr iŋ ād) *noun* A small electronic device that makes it easier for a person to hear sounds. *Many hearing aids are small enough to fit into the ear canal.*

a	cat	e	net	îr	gear	u	cup	u̇	look, pull	*th*	**this**	ə	alive,
ā	day, lake	ē	seed	o	hot	ū	fuse	oi	soil	hw	**wheel**		comet,
ä	father	i	fit	ō	cold	ûr	fur, bird	ou	**out**	zh	measure		acid, atom,
âr	dare	ī	pine	ô	paw	ü	tool, rule	th	thin	ŋ	wing		focus

heart (härt) *noun* The organ that pumps blood throughout the body. *The human heart is a special kind of muscle about the size of a fist.*

heat (hēt) *noun* A form of energy that comes from the movement of the particles in a substance. *Heat sources for buildings include oil, gas, electricity, and sunlight.* **heat** *verb*

height (hīt) *noun* The measure of an object from top to bottom. *The height of the pine tree is 10 meters, or almost 40 feet.*

he·li·um (hē' lē əm) *noun* A very light element that is a gas. *Helium is often used in toy balloons.*

hel·per T-cells (hel' pər tē' selz) *noun, plural* Immune system cells that help destroy substances that make people sick. *Helper T-cells fight infections.*

her·bi·vore (hûr' bə vôr) *noun* An animal that eats only plants. *Cows and horses are herbivores.* **herbivorous** *adjective*

he·red·i·ty (hər ed' i tē) *noun* The passing of traits from parent to offspring. *Eye color is determined by heredity.* **hereditary** *adjective*

hi·ber·nate (hī' bər nāt) *verb* To spend the winter asleep or inactive. *Bears, turtles, and frogs all hibernate.*

hi·ber·na·tion (hī bər nā' shən) *noun* A long period when an animal is not active and all its body systems slow down. *Hibernation lets animals survive very cold weather and lack of food.*

high blood pres·sure (hī blud presh' ər) *noun* A condition in which the heart has to work extra hard to keep blood moving through the body. *High blood pressure is a major cause of strokes and heart disease.*

high-pres·sure ar·ea (hī' presh ər âr' ē ə) *noun* A region where air is sinking, causing air pressure to be high. *High-pressure areas usually have fair weather.*

high tide (hī tīd) *noun* The time when the tide reaches its highest level. *High tide occurs about every 12½ hours at any given point on the coast.*

hinge joint (hinj joint) *noun* A joint where bones can move in only one direction. *Elbows and knees are hinge joints.*

ho·ri·zon (hə rī' zən) *noun* 1. The place where Earth and sky appear to meet. *The Moon rose over the horizon at dusk.* 2. A layer of soil different from the layers above and below it. *Halfway up the cliff was a dark horizon between two limestone layers.* **horizontal** *adjective*

host (hōst) *noun* A plant or animal on which a parasite lives and feeds. *Fleas live on hosts such as dogs, cats, and mice.*

hu·mid·i·ty (hū mid′ i tē) *noun*
A measure of how much water vapor is in the air. *When the humidity is high, the air feels damp.* **humid** *adjective*

hu·mus (hū′ məs) *noun* Rich, dark material in the soil made of decayed plants and animals. *Humus helps make soil fertile.*

hur·ri·cane (hûr′ ə kān) *noun* A huge, powerful storm with strong winds and heavy rains. *Hurricanes whirl around a calm, clear center, called the eye.*

hy·dro·elec·tric·i·ty
(hī drō i lek tris′ ə tē) *noun* Electricity produced by falling water. *The Hoover Dam uses the flow of the Colorado River to make hydroelectricity.*

hy·grom·e·ter (hī grom′ i tər) *noun* An instrument for measuring moisture in the air. *We use hygrometers to help forecast the weather.*

ice (īs) *noun* Water in its solid state. *Water turns to ice when it is cooled below its freezing point, 0°C (32°F).*

Ice Age (īs āj) *noun* A period of time when a large part of Earth was covered with glacial ice. *The last Ice Age ended about 10,000 years ago.*

ice cap (īs kap) *noun* A thick sheet of ice covering a large land area. *Most of Baffin Island in Canada is under an ice cap.*

i·den·ti·fy (ī den′ tə fī) *verb* To recognize a plant, animal, or other specimen as a certain kind by comparing it with other examples. *The students identified the insects they had collected in the field.* **identification** *noun*

ig·ne·ous (ig′ ne əs) *adjective* A kind of rock formed when melted rock material cools. *Granite and basalt are types of igneous rock.*

il·le·gal drug (i lē′ gəl drug) *noun* Substance harmful to the body and against the law to use. *Heroin and cocaine are examples of illegal drugs.*

a	cat	e	net	îr	gear	u	cup	u̇	look, pull	*th*	*this*	ə	alive,
ā	day, lake	ē	seed	o	hot	ū	fuse	oi	soil	hw	**wheel**		comet,
ä	father	i	fit	ō	cold	ûr	fur, bird	ou	**out**	zh	measure		acid, atom,
âr	dare	ī	pine	ô	paw	ü	tool, rule	th	**thin**	ŋ	wing		focus

im·mov·able joint (i mü′ və bəl joint) *noun* A joint where the bones do not move. *Immovable joints are found in the skull.*

im·mune sys·tem (i mūn′ sis′ təm) *noun* The system that protects the body against disease and infection. *White blood cells are part of the human immune system.*

im·mu·ni·ty (i mū′ ni tē) *noun* Protection from disease. *Infants get vaccinations that give them immunity from measles and mumps.* **immune** *adjective*

im·print (im′ print) *noun* A shallow mark made by stamping or pressing an object onto a soft surface. *You can make an imprint by pushing a shell into clay.*

im·pulse (im′ puls) *noun* A message sent through the body along nerves. *Impulses carry information from your sense organs to your brain.*

in·can·des·cent (in kən des′ ənt) *adjective* Giving off light when electric current flows through a filament. *The desk lamp has a 60-watt incandescent light bulb.*

in·cin·er·a·tion (in sin ər ā′ shən) *noun* The process of burning to ashes. *Incineration of wood takes a very hot fire.* **incinerate** *verb*

in·ci·sor (in sī′ zər) *noun* One of eight front teeth used for cutting. *A child's first permanent teeth are usually the incisors.*

in·clined plane (in′ klīnd plān) *noun* A simple machine made of a flat surface with one end higher than the other. *A ramp is an inclined plane.*

in·di·ges·tion (in di jes′ chən) *noun* Difficulty digesting food. *People feel sick when they have indigestion.*

in·er·tia (in ûr′ shə) *noun* An object's tendency to either keep moving or stay still unless a force acts on it. *Inertia is why we need to use brakes to stop a speeding car.*

in·ex·haust·i·ble re·source (in eg zôs′ tə bəl rē′ sôrs) *noun* An energy source that cannot be used up. *The Sun and wind are inexhaustible resources.*

in·fer (in fûr′) *verb* To draw a conclusion based on evidence. *We inferred that the white substance was salt.* **inference** *noun*

in·fra·red ra·di·a·tion (in frə red′ rā dē ā′ shən) *noun* Light energy that transfers heat. *We feel infrared radiation as heat, but we cannot see it.*

in·hale (in hāl′) *verb* To breathe in. *When you inhale, your lungs fill with air.*

in·her·it (in her′ it) *verb* To receive traits from biological parents. *Children inherit hair color and eye color from their parents.*

in·her·it·ed be·hav·ior (in her′ it əd bi hāv′ yər) *noun* A way of acting that an animal is born with. *Flying is an inherited behavior of birds.*

inherited trait (in her′ it əd trāt) *noun* A characteristic passed from parents to offspring through genes. *Tall parents often have tall children because height is partly an inherited trait.*

in·ner core (in′ ər kôr) *noun* The center region of the inside of Earth. *The inner core is a very hot mass of solid iron and nickel.*

in·ner ear (in′ ər îr) *noun* The innermost part of the ear, where sound vibrations are turned into sound signals. *Sounds travel from the inner ear to the brain along the auditory nerve.*

in·ner planets (in′ ər plan′ its) *noun, plural* The planets in the solar system closest to the Sun. *Mercury, Venus, Earth, and Mars are the inner planets.*

in·sect (in′ sekt) *noun* A small animal with six legs, one or two pairs of wings, and three main sections to its body. *Insects, such as dragonflies, have an exoskeleton and no backbone.*

in·stinct (in′ stiŋkt) *noun* Something that an animal knows how to do without being taught. *Spiders spin webs by instinct.* **instinctive** *adjective*

in·su·la·tor (in′ sə lā tər) *noun* A material that blocks the flow or escape of heat, sound, or electricity. *Electrical wires are usually wrapped with an insulator.* **insulate** *verb*

in·ter·act (in tər akt′) *verb* To have an effect on each other. *The plants and animals that live in and around the pond interact with one another.*

in·ver·te·brate (in vûr′ tə brāt) *noun* An animal without a backbone. *Insects, worms, and snails are invertebrates.*

in·vol·un·tar·y mus·cle (in vol′ ən ter ē mus′ əl) *noun* A muscle that does its work automatically. *The heart is made of involuntary muscle.*

i·ris (ī′ ris) *noun* The colored part of the eye. *The iris is a muscle that controls how much light enters the pupil.*

i·ron (ī′ ərn) *noun* An element that is a strong, hard metal. *Iron is used to make fences and outdoor furniture.*

ir·ri·ga·tion (îr ə gā′ shən) *noun* A way to get water into the soil for crops. *Farmers often use canals and ditches for irrigation.* **irrigate** *verb*

a	cat	e	net	îr	gear	u	cup	u̇	look, pull	th	this	ə	alive,
ā	day, lake	ē	seed	o	hot	ū	fuse	oi	soil	hw	wheel		comet,
ä	father	i	fit	ō	cold	ûr	fur, bird	ou	out	zh	measure		acid, atom,
âr	dare	ī	pine	ô	paw	ü	tool, rule	th	thin	ŋ	wing		focus

jet stream (jet strēm) *noun* A very strong current of air high above Earth. *Jet streams move at high speed from west to east.*

joint (joint) *noun* A place where bones meet in a way that allows movement. *Knees, hips, shoulders, and elbows are joints.*

Ju·pi·ter (jü' pi tər) *noun* The fifth planet from the Sun in our solar system. *Jupiter is the largest of our Sun's planets.*

kid·ney (kid' nē) *noun* An organ that removes waste from the blood. *Humans have two kidneys, located in the lower back.*

ki·lo·gram (kil' ə gram) *noun* A metric unit for measuring mass. *One thousand grams equal 1 kilogram.*

ki·lo·me·ter (kil' ə mē tər) *noun* A metric unit for measuring distance. *One thousand meters equal 1 kilometer.*

ki·net·ic en·er·gy (kə net' ik en' ər jē) *noun* The energy of motion. *A galloping horse and a falling rock have kinetic energy.*

king·dom (kiŋ' dəm) *noun* One of the five main groups of living things. *The five kingdoms are animals, plants, protists, monerans, and fungi.*

land·fill (land' fil) *noun* A place where garbage is dumped and covered with earth. *Hazardous waste should never be put in landfills.*

land·form (land' fôrm) *noun* A natural shape or feature of Earth's surface. *Mountains and valleys are landforms.*

large in·tes·tine (lärj in tes' tin) *noun* The wide lower end of the digestive system. *The large intestine absorbs water from undigested food.*

lar·va (lär' və) *noun* The wormlike early stage in the insect life cycle, between egg and pupa. *A caterpillar is the larva of a butterfly or moth.* **larvae** plural

lat·i·tude (lat' i tüd) *noun* The distance north or south of the equator on Earth's surface, measured in degrees. *Halfway between the equator and the North Pole is 45˚ north latitude.*

la·va (lä′ və) *noun* The hot liquid rock that pours out of an erupting volcano. *When lava cools, it hardens and forms solid igneous rock.*

lava flow (lä′ və flō) *noun* A stream of molten lava. *The lava flow from the island volcano cooled when it reached the ocean.*

leaf (lēf) *noun*
The flat green part of a plant or tree that grows from a stem. *Leaves make food for the plant.*
leaves *plural*

learned be·hav·ior (lürnd bi hāv′ yər) *noun* Behavior that a person or animal is taught or learns from experience. *Riding a bicycle is a learned behavior.*

length (leŋkth) *noun* The measure of an object from one end to the other. *The length of the canoe is about 5 meters (16 feet).*

lens (lenz) *noun* 1. A curved piece of glass or plastic that bends light rays passing through it. *Different kinds of lenses can make an object look smaller or larger than it is.* 2. The curved, transparent part of the eye that focuses light on the retina. *Light enters the eye through the pupil and passes through the lens.*

lev·er (lev′ ər) *noun* A rigid bar that turns around a point or fulcrum. *A lever is a simple machine for lifting loads.*

life cy·cle (līf sī′ kəl) *noun* The stages in an organism's life. *The life cycle of many insects includes the egg, larva, pupa, and adult stages.*

life span (līf span) *noun* The length of time an organism is expected to live. *The life span of the giant tortoise is about 170 years.*

lig·a·ment (lig′ ə mənt) *noun* A tissue that joins bones together. *Strong ligaments connect the leg bones at the knee.*

light·ning (līt′ niŋ) *noun* A flash of light in the sky when electricity moves between clouds or between clouds and the ground. *Summer storms often have lightning and thunder.*

light-year (līt′ yîr) *noun* The distance light travels in one year, almost 9.5 trillion kilometers (about 6 trillion miles). *Scientists measure distance in space in light years.*

line graph (līn graf) *noun* A graph that shows how something changes over time. *A line graph is a useful tool for showing temperature changes.*

a	cat	e	net	îr	gear	u	cup	u̇	look, pull	th	this	ə	alive,
ā	day, lake	ē	seed	o	hot	ū	fuse	oi	soil	hw	wheel		comet,
ä	father	i	fit	ō	cold	ûr	fur, bird	ou	out	zh	measure		acid, atom,
âr	dare	ī	pine	ô	paw	ü	tool, rule	th	thin	ŋ	wing		focus

line of force (līn əv fôrs) *noun* A line in a magnetic field. *Iron filings form a pattern along the lines of force around a magnet.*

liq·uid (lik' wid) *noun* Matter that has a set volume but no set shape. *Liquids, such as water, milk, and juice, take the shape of their containers and flow easily.*

li·ter (lē' tər) *noun* The basic unit of volume in the metric system. *One liter equals 1,000 milliliters.*

lit·ter (lit' ər) 1. *noun* Trash or garbage scattered around carelessly. *Litter is a major pollution problem.* 2. *verb* To throw away trash carelessly. *Please do not litter the schoolyard.*

load (lōd) *noun* The object that a simple machine moves. *Push down on that lever to lift the load at the other end.*

loam (lōm) *noun* Rich soil that is made up of clay, sand, silt, and humus. *Most plants grow well in loam.*

lode·stone (lōd' stōn) *noun* A hard, black, naturally magnetic rock; sometimes spelled loadstone. *Lodestones attract iron.*

lon·gi·tude (lon' ji tüd) *noun* The distance east or west of the prime meridian on Earth's surface, measured in degrees. *The prime meridian passes through Greenwich, England, located at 0° longitude.*

low-pres·sure ar·ea (lō' presh ər âr' ē ə) *noun* A region where air is rising, causing air pressure to be lower. *Low-pressure areas often have stormy weather.*

lu·nar (lü' nər) *adjective* Having to do with the Moon. *The lunar cycle from new Moon to new Moon lasts about 28 days.*

lunar e·clipse (lü' nər ē klips') *noun* The darkening or hiding of the Moon when it passes through Earth's shadow. *The Earth is between the Moon and the Sun during a lunar eclipse.*

lungs (luŋz) *noun, plural* The two baglike organs in the chest used in breathing. *The lungs are the main organs of the respiratory system.*

lus·ter (lus' tər) *noun* The way a mineral reflects light. *The gold medal has a beautiful luster.*

ma·chine (mə shēn') *noun* A device that uses energy and does work. *A lawnmower is a machine that cuts grass.*

mag·ma (mag' mə) *noun* Melted rock inside Earth. *When magma reaches the surface, it is called lava.*

magma cham·ber (mag' mə chām' bər) *noun* An underground area filled with melted rock. *Lava comes out of volcanoes from deep magma chambers.*

mag·net (mag′ nət) *noun* A piece of metal that attracts iron or steel. *Some magnets are shaped like horseshoes.*

mag·net·ic field (mag nət′ ik fēld) *noun* The space around a magnet where the force of the magnet can act. *A nail inside a magnetic field will be pulled to the magnet.*

magnetic poles (mag nət′ ik pōlz) *noun, plural* The parts of a magnet where the force is strongest. *The magnetic poles are labeled north and south.*

mag·net·ism (mag′ nə tiz əm) *noun* The force of a magnet that draws some objects to it. *The magnetism of Earth is what makes compasses work.*

mag·ni·fi·er (mag′ nə fī ər) *noun* A convex lens that makes an object appear larger than it is. *You can use a magnifier to see the ridges on your fingerprint.* **magnify** *verb*

mal·le·a·ble (mal′ ē ə bəl) *adjective* Easily hammered or molded into a different shape. *Tin is a malleable metal.*

mam·mal (mam′ əl) *noun* A warm-blooded animal that has fur or hair and produces milk for its young. *Monkeys, mice, and whales are mammals.*

mam·moth (mam′ əth) *noun* A large elephant-like mammal that lived long ago and is now extinct. *Mammoths had long curved tusks and shaggy hair.*

man·di·ble (man′ də bəl) *noun* The mouth part of some insects that helps it hold or bite food. *The beetle used its mandible to hold and chew the leaf.*

man·tle (man′ təl) *noun* The thickest layer of Earth, made of molten rock under pressure. *Earth's mantle is located between the crust and the core.*

Mar·i·an·a Trench (mâr ē an′ ə trench) *noun* The deepest part of the world ocean, located in the Pacific near the Philippines. *The Mariana Trench is so deep that no light reaches the bottom.*

mar·row (mâr′ ō) *noun* The soft tissue inside bones. *Bone marrow produces red and white blood cells.*

Mars (märz) *noun* The fourth planet from the Sun in our solar system. *Mars is sometimes called the red planet.*

mass (mas) *noun* The amount of matter in an object. *Mass can be measured in grams using a balance.*

a	cat	e	net	îr	gear	u	cup	ů	look, pull	*th*	**this**	ə	alive,
ā	day, lake	ē	seed	o	hot	ū	fuse	oi	soil	hw	**wheel**		comet,
ä	father	i	fit	ō	cold	ûr	fur, bird	ou	**out**	zh	measure		acid, atom,
âr	dare	ī	pine	ô	paw	ü	tool, rule	th	**thin**	ŋ	wi**ng**		focus

mass ex·tinc·tion (mas ek stiŋk' shən) *noun* The dying out of many species at the same time. *About 65 million years ago, something caused a mass extinction on Earth.*

mat·ter (mat' ər) *noun* Anything that takes up space and has mass. *Matter has three basic states: solid, liquid, and gas.*

maze (māz) *noun* Paths or lines laid out like a puzzle to find your way through. *It is fun to figure out a maze.*

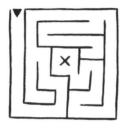

me·chan·i·cal en·er·gy (mə kan' i kəl en' ər jē) *noun* The energy of an object that is moving or that can move. *Skateboards and spinning tops have mechanical energy.*

me·di·an (mē' dē ən) *noun* The middle number in a set of numbers that are in order. *In the number set 2, 4, 6, 8, 10, the median is 6.*

melt (melt) *verb* To change from a solid to a liquid by heating. *The snow will melt on a mild winter day.*

melt·ing point (mel' tiŋ point) *noun* The temperature at which a solid becomes a liquid. *The melting point of ice is 0° Celsius (32° Fahrenheit).*

mem·brane (mem' brān) *noun* A thin tissue that covers a part of an animal or plant. *The eardrum is a membrane.*

mer·cu·ry (mûr' kū rē) *noun* A heavy, silvery metal that is an element. *Most metals are solids, but mercury is a liquid at room temperature.*

Mer·cu·ry (mûr' kū rē) *noun* The nearest planet to the Sun in our solar system. *Mercury is less than half the size of Earth and orbits the Sun in 88 Earth days.*

Mes·o·zo·ic era (mez ə zō' ik îr' ə) *noun* The period in Earth's history between 245 and 65 million years ago. *Dinosaurs lived on Earth during the Mesozoic era.*

met·al (met' əl) *noun* An element that is usually hard and shiny in its pure form. *We use the metals gold, silver, copper, and tin almost every day.*

met·al·lur·gy (mə' təl ər jē) *noun* The study of metals. *Finding ways to remove metals from rocks is part of metallurgy.*

met·a·mor·phic rock (met ə môr' fik rok) *noun* A type of rock formed when another rock is changed by heat or pressure. *Slate, schist, and gneiss are kinds of metamorphic rocks.*

met·a·mor·pho·sis (met ə môr' fə sis) *noun* A change in form that some animals go through in their life cycles. *The change of a tadpole into a frog is an example of metamorphosis.*

me·te·or (mē' tē ər) *noun* A small piece of rock or metal in space that burns up when it enters Earth's atmosphere. *A meteor is often called a shooting star.*

me·te·or·ite (mē' tē ər īt) *noun* A meteor that reaches the surface of a planet or moon without completely burning up. *A meteorite can form a crater when it lands.*

me·te·or·oid (mē′ tē ər oid) *noun*
A small rocky or metal object that orbits the Sun. *Meteoroids can be as small as specks of dust.*

me·te·o·rol·o·gist (mēt ē ər ôl′ ə jist) *noun* A scientist who studies the weather. *Meteorologists use data from many instruments to make weather forecasts.*

me·te·o·rol·o·gy (met ē ər ôl′ ə jē) *noun* The study of Earth's climate and weather. *Satellite photos are used to observe weather patterns in meteorology.*

me·ter (mē′ tər) *noun* The basic unit of length in the metric system. *One meter equals 100 centimeters, or about 39 inches.*

met·ric sys·tem (me′ trik sis′ təm) *noun* A system of measurement based on units of 10. *Most countries around the world use the metric system to measure and count.*

mi·cro·or·gan·ism (mī kro ôr′ gə niz əm) *noun* A living thing too small to be seen with the naked eye. *Bacteria and yeasts are microorganisms.*

mi·cro·phone (mī′ krə fōn) *noun* A device that turns sound waves into electrical waves. *Microphones are used to make sounds louder or to record them.*

mi·cro·scope (mi′ krə skōp) *noun* An instrument that uses lenses to make very small objects look larger. *Scientists use microscopes to study cells.* **microscopic** *adjective*

mid·dle ear (mid′ əl îr) *noun* The section of the ear in which small bones transmit sound wave vibrations. *The middle ear is between the eardrum and the cochlea.*

mi·grate (mī′ grāt) *verb* To move from one area to another when seasons change. *Flocks of geese migrate to warm regions for winter.*

mi·gra·tion (mī grā′ shən) *noun* The regular long-distance movement of animals from one region to another. *Migration of reindeer herds begins a few months after the young are born.*

miles per hour (mīlz pər our) *noun* A measure of speed, which includes the distance traveled and the time it took. *The speed limit on many highways is 65 miles per hour (105 kilometers per hour).*

Milk·y Way (mil′ kē wā) *noun* The name of the galaxy that includes our solar system and billions of stars. *The Milky Way looks like a band of fuzzy light across the night sky.*

a	cat	e	net	îr	gear	u	cup	ů	look, pull	th	this	ə	alive,
ā	day, lake	ē	seed	o	hot	ū	fuse	oi	soil	hw	wheel		comet,
ä	father	i	fit	ō	cold	ûr	fur, bird	ou	out	zh	measure		acid, atom,
âr	dare	ī	pine	ô	paw	ü	tool, rule	th	thin	ŋ	wing		focus

mil·li·gram (mil′ i gram) *noun* A unit of mass in the metric system. *One thousand milligrams equal 1 gram.*

mil·li·li·ter (mil′ ə lē tər) *noun* A unit of volume or capacity in the metric system. *One thousand milliliters equal 1 liter.*

mil·li·me·ter (mil′ ə mē tər) *noun* A unit of length in the metric system. *One thousand millimeters equal 1 meter.*

mim·ic·ry (mim′ i krē) *noun* A color, shape, or design that makes an animal look like another animal or object. *Mimicry helps animals hide from danger.* **mimic** *verb*

min·er·al (min′ ər əl) *noun* A solid substance formed in nature that has never been alive. *Gold, mica, and quartz are examples of minerals.*

mis·use (mis ūz′) *verb* To use something in an unsafe or wrong way. *Be very careful not to misuse the chemicals in science class.*

mix·ture (miks′ chər) *noun* Different kinds of matter mixed together but keeping their own properties. *Air is a mixture of gases.* **mix** *verb*

mode (mōd) *noun* The number that appears most often in a set of numbers. *In the number set 1, 4, 5, 5, 8, 8, 8, 9, the mode is 8.*

mod·el (mod′ əl) *noun* Something that shows how an object or system works. *The solar system model shows the orbits of the planets around the Sun.*

Mohs′ scale (mōz scāl) *noun* A way to describe the hardness of a mineral. *A diamond, the hardest mineral, measures 10 on the Mohs′ scale.*

mold (mōld) *noun* A kind of fungus that grows on plant or animal matter. *Mold began to grow on the stale bread.*

mold fos·sil (mōld fos′ əl) *noun* An imprint left in a rock by a plant or animal. *A mold fossil shows the outside parts of a dead organism, such as the fins and scales of a prehistoric fish.*

mol·e·cule (mol′ ə kūl) *noun* The smallest unit of a compound. *A molecule is made up of more than one atom.*

mol·lusk (mol′ əsk) *noun* An animal with a soft body and a hard outer shell. *A clam and a nautilus are both mollusks.*

molt (mōlt) *verb* To shed an outer layer before getting a new covering of skin, hair, feathers, or shell. *Insects, snakes, and many other animals molt in order to grow.*

mol·ten (mōl′ tən) *adjective* Made liquid by heating. *Magma is molten rock.*

mon·er·an (mon′ ər an) *noun* A one-celled organism without a nucleus. *Bacteria and blue-green algae are monerans.*

mon·o·cot seed (mon′ ə kot sēd) *noun* A plant seed that produces one cotyledon, or seed leaf. *Corn is an example of a monocot seed.*

moon (mün) *noun* A natural satellite that orbits a planet. *Mars has 2 moons, and Jupiter has 39 moons.*

Moon (mün) *noun* The natural satellite that orbits Earth. *The Moon is Earth's nearest neighbor in space.*

mo·raine (mə rān') *noun* Soil and rocks carried by a glacier and left behind at the edge of the glacier as it melts. *Cape Cod in Massachusetts is a moraine.*

mo·tion (mō' shən) *noun* Movement, or a change of position. *Speed and direction are ways to describe motion.*

moun·tain (moun' tən) *noun* A landform that is much higher than the land around it. *Everest is the highest mountain on Earth.*

mov·a·ble joint (müv' ə bəl joint) *noun* A place where bones connect in ways that allow body parts to move. *The hinge joints in our elbows are movable joints.*

mov·a·ble pul·ley (müv' ə bəl púl' ē) *noun* A pulley that is attached to the load and moves in the same direction as the force. *A movable pulley reduces the force needed to do work.*

mus·cle (mus' əl) *noun* The body tissue that connects to the bones and moves them by getting longer and shorter. *Biceps and triceps are arm muscles.*

mus·cu·lar sys·tem (mus' kyə lər sis' təm) *noun* The body system that produces movement. *The muscular system includes muscles, tendons, and ligaments.*

Na·tion·al Weath·er Ser·vice (nash' ə nəl weth' ər sûr' vəs) *noun* The government organization that provides weather data, forecasts, and warnings for the United States. *The National Weather Service issues hurricane and tornado warnings.*

na·tur·al gas (nach' ər əl gas) *noun* A gas found underground, formed millions of years ago. *Natural gas is used as fuel for heating and cooling.*

na·ture (nā' chər) *noun* The physical world and everything in it not made by humans. *Nature includes plants, animals, the weather, and landforms.*
natural *adjective*

nec·tar (nek' tər) *noun* The sweet liquid inside some flowers. *Bees collect nectar and turn it into honey.*

a	cat	e	net	îr	gear	u	cup	ů	look, pull	*th*	**this**	ə	alive,
ā	day, lake	ē	seed	o	hot	ū	fuse	oi	soil	hw	**wheel**		comet,
ä	father	i	fit	ō	cold	ûr	fur, bird	ou	out	zh	measure		acid, atom,
âr	dare	ī	pine	ô	paw	ü	tool, rule	th	thin	ŋ	wing		focus

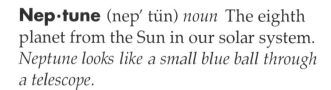

Nep·tune (nep′ tün) *noun* The eighth planet from the Sun in our solar system. *Neptune looks like a small blue ball through a telescope.*

nerve (nûrv) *noun* A bundle of cells that carries signals throughout the body. *Nerves carry messages to and from the brain.*

nerve cell (nûrv sel) *noun* A cell that makes up a nerve; also called a neuron. *Nerve cells have a star-shaped body and long "tail."*

nerve end·ing (nûrv end′ iŋ) *noun* The part of a nerve cell where signals are passed to the next nerve. *He damaged nerve endings when he cut the end of his finger.*

ner·vous sys·tem (nûr′ vəs sis′ təm) *noun* The organ system that controls all the other organ systems. *The brain, spinal cord, and nerves make up the nervous system.*

neu·ron (nûr′ on) *noun* A nerve cell. *Neurons in the skin are used for our sense of touch.*

neu·tral (nü′ trəl) *adjective* Neither an acid nor a base. *The stream water we tested was neutral.*

neu·tron (nü′ tron) *noun* One of the tiny particles in the nucleus of an atom. *A neutron has no electric charge.*

new·ton (nü′ tən) *noun* A unit for measuring force. *The newton is named for Sir Isaac Newton, an important scientist.*

niche (nich) *noun* The role or position of an organism in its habitat. *An animal's niche depends on where it lives, what it eats, and what its enemies are.*

ni·tro·gen (nī′ trə jən) *noun* An element that is a colorless, odorless, tasteless gas. *Nitrogen makes up more than three-fourths of Earth's atmosphere.*

non·liv·ing (non liv′ iŋ) *adjective* Never having been alive. *A spoon is an example of a nonliving object.*

non·met·al (non met′ əl) *noun* An element that is not a metal. *Carbon, oxygen, and helium are nonmetals.*

non·re·new·a·ble re·source (non rē nü′ ə bəl rē′ sôrs) *noun* A natural resource or source of energy that can be used up. *Coal and petroleum are examples of nonrenewable resources.*

non·vas·cu·lar plant (non vas′ kyə lər plant) *noun* A plant that has no tubes for carrying water and sap to its parts. *Mosses are nonvascular plants.*

North·ern Hem·i·sphere (nôr′ thərn hem′ i sfîr) *noun* The half of Earth that is north of the equator. *The United States is in the Northern Hemisphere.*

North Pole (nôrth pōl) *noun* The most northern point on Earth. *The North Pole is located at the northern tip of Earth's axis.*

nu·cle·ar en·er·gy (nü′ klē ər en′ ər jē) *noun* Energy from the nucleus of an atom that is released by fission or fusion. *Nuclear energy is the most powerful form of energy.*

nuclear fis·sion (nü′ klē ər fish′ ən) *noun* Splitting the nucleus of an atom, usually uranium. *Nuclear fission gives off huge amounts of energy.*

nuclear fu·sion (nü′ klē ər fū′ zhən) *noun* Combining the nuclei of more that one atom, usually hydrogen. *Nuclear fusion is the source of the Sun's energy.*

nu·cle·us (nü′ klē əs) *noun* 1. The central part of a living cell. *The nucleus controls the activities of the cell.* 2. The center of an atom. *Electrons orbit the nucleus of an atom.* **nuclear** *adjective* **nuclei** *plural*

nu·tri·ent (nü′ trē ənt) *noun* A substance needed by an organism for health and growth. *The food we eat contains nutrients.* **nutritious** *adjective*

nymph (nimf) *noun* A stage in the life cycle of some insects between egg and adult. *A dragonfly nymph looks much like the adult, but smaller and without wings.*

o·cean (ō′ shən) *noun* The large body of salt water that covers about 70 percent of Earth. *The world ocean is divided into the Pacific, Atlantic, Indian, Arctic, and Antarctic Oceans.*

oil (oil) *noun* A thick, greasy substance that burns easily and does not mix with water. *Oils are used as fuels and lubricants.*

ol·fac·to·ry nerve (ōl fak′ tər e nûrv) *noun* The nerve that carries information about smells from the nose to the brain. *The olfactory nerve is located on the roof of the nasal cavity.*

om·ni·vore (om′ ni vôr) *noun* An animal that eats both plants and other animals. *Humans who eat meat and vegetables are omnivores.* **omnivorous** *adjective*

o·paque (o pāk′) *adjective* Able to block all light. *Metal is opaque, but glass is not.*

o·pen cir·cuit (ō′ pən sûr′ kit) *noun* A broken or incomplete electrical path. *Electric current cannot flow through an open circuit.*

a	cat	e	net	îr	gear	u	cup	u̇	look, pull	*th*	**this**	ə alive,
ā	day, lake	ē	seed	o	hot	ū	fuse	oi	soil	hw	**wheel**	comet,
ä	father	i	fit	ō	cold	ûr	fur, bird	ou	**out**	zh	measure	acid, atom,
âr	dare	ī	pine	ô	paw	ü	tool, rule	th	thin	ŋ	wi**ng**	focus

open o·cean (ō′ pən ō′ shən) *noun*
The deep part of the ocean, far from
shore. *Whales live in the open ocean.*

op·tic nerve (op′ tik nûrv) *noun*
The nerve that carries information about
light from the eye to the brain. *The optic
nerve is located at the back of the retina.*

or·bit (ôr′ bit) *noun* The path of an
object as it travels
around a larger object.
*Earth's orbit around the
Sun is in the shape of an
ellipse.* **orbit** *verb*

or·der (ôr′ dər) *noun* In classification of
living things, a smaller group within a
class. *The mammal class is divided into
more than 20 orders.*

ore (ôr) *noun* A rock that contains a
high percentage of a metal. *Iron must be
removed from iron ore before it can be made
into steel.*

or·gan (ôr′ gən) *noun* A group of
tissues that work together to do a job in
the body. *The brain, stomach, and heart are
examples of organs.*

or·gan·ism (ôr′ gə niz əm) *noun*
A living thing. *Organisms are an
important part of nature.*

or·gan sys·tem (ôr′ gən sis′ təm) *noun*
A group of organs that work together.
*Muscles make up the organ system that
works so we can move.*

out·er core (ou′ tər kôr) *noun* A liquid
layer of Earth just below the mantle. *The
outer core is about 2,200 kilometers (about
1,370 miles) thick and surrounds the solid
inner core.*

out·er ear (ou′ tər îr) *noun* The part of
the ear that sticks out from the head. *The
outer ear captures sound waves and funnels
them into the ear canal.*

out·er plan·ets (ou′ tər plan′ its) *noun,
plural* The planets in the solar system
that are farthest from the Sun. *Jupiter,
Saturn, Neptune, Uranus, and Pluto are the
outer planets.*

out·wash plain (out′ wâsh plān) *noun*
Gravel and sand carried by streams from
melting glaciers. *Outwash plains are left
behind over large areas as glaciers melt.*

o·var·y (ō′ vər ē) *noun* The female
organ that produces eggs. *The ovary of a
flower forms the fruit.*

o·vule (ō′ vūl) *noun* The part of a plant
that forms a seed. *The ovule is inside the
ovary.*

ox·y·gen (ok′ sə jən) *noun* The element
in air that is needed by most organisms
to live. *Animals take oxygen out of the air
when they breathe.*

o·zone lay·er (ō′ zōn lā′ ər) *noun*
A layer of Earth's atmosphere containing
a lot of ozone, a form of oxygen. *The
ozone layer blocks much of the Sun's harmful
rays from Earth's surface.*

41

pa·le·on·tol·o·gist (pā lē ən tol' ə jist) *noun* A scientist who studies the remains of living things from the ancient past. *Paleontologists find fossils and look for clues to prehistoric life.* **paleontology** *noun*

Pa·le·o·zo·ic era (pā lē ə zō' ik îr' ə) *noun* The period of Earth's history from about 545 million to about 245 million years ago. *Amphibians, fish, and insects first appeared on Earth during the Paleozoic era.*

Pan·gae·a (pan jē' ə) *noun* The huge land mass that was the only continent on Earth millions of years ago. *Pangaea began to break up into today's continents about 200 million years ago.*

par·al·lel cir·cuit (pâr' ə lel sûr' kit) *noun* An electrical circuit split into different paths. *Each device in a parallel circuit is connected separately to the power source.*

par·a·site (pâr' ə sīt) *noun* An organism that lives in or on another organism. *Fleas are parasites that often live on dogs and cats.*

pas·sive smoke (pas' iv smōk) *noun* Smoke that is inhaled by someone other than the smoker. *Passive smoke can be harmful to nonsmokers.*

pat·tern (pat' ərn) *noun* A repeating design or series of events. *The pattern of a zebra's stripes helps it blend in with tall grasses.*

peat (pēt) *noun* A soft material made from decayed plants. *Peat is the first stage in the creation of coal.*

pen·i·cil·lin (pen ə sil' in) *noun* A type of antibiotic that kills disease-causing bacteria. *Penicillin was first made in 1928 from a type of mold.*

pen·in·su·la (pə nin' sə lə) *noun* A long piece of land almost completely surrounded by water. *A peninsula is almost an island.*

pe·ri·od·ic (pîr ē od' ik) *adjective* Repeating in a pattern. *Periodic rain over a short time can cause flooding.*

per·ish (per' ish) *verb* To die, or to be destroyed or spoiled. *When all the animals of one kind perish, the animal is extinct.* **perishable** *adjective*

a	cat	e	net	îr	gear	u	cup	u̇	look, pull	th	this	ə	alive,
ā	day, lake	ē	seed	o	hot	ū	fuse	oi	soil	hw	wheel		comet,
ä	father	i	fit	ō	cold	ûr	fur, bird	ou	out	zh	measure		acid, atom,
âr	dare	ī	pine	ô	paw	ü	tool, rule	th	thin	ŋ	wing		focus

per·me·a·bil·i·ty (pûr mē ə bil′ i tē)
noun The rate at which water can pass
through a substance. *Sandy soils have high
permeability.* **permeable** *adjective*

pes·ti·cide (pes′ tə sīd) *noun*
A chemical used to kills insects, weeds,
or other pests. *Homeowners sometimes use
pesticides to get rid of ants.*

pet·al (pet′ əl) *noun* The colored or
showy part of many flowers. *Red petals
can attract hummingbirds to a flower.*

pet·ri·fied (pet′ rə fīd) *adjective*
Preserved by being turned to stone.
*When minerals seep into the cells of dead,
fallen trees, the trunks become petrified
wood.* **petrify** *verb*

pet·ri·fied fos·sil (pet′ rə fīd fos′ əl)
noun Plant or animal remains whose
cells have been replaced with minerals.

pe·tro·leum (pə trō′ lē əm) *noun*
A thick, dark, liquid fossil fuel. *Petroleum
is crude oil.*

phar·ma·cist (fär′ mə sist) *noun*
A person who prepares and sells
medicines according to a doctor's
instructions. *Pharmacists must be well
trained in chemistry.*

phas·es (fāz əz) *noun, plural* The
different shapes the Moon appears to
have when seen from Earth. *The Moon
goes through its cycle of phases about every
4 weeks.*

pho·to·syn·the·sis (fō to sin′ thə sis)
noun The process by which green plants
use sunlight to change carbon dioxide
and water into food and oxygen.
*Photosynthesis takes place in a plant's
leaves.*

pH scale (pē āch′ skāl) *noun* A ranking
of acid, neutral, and base substances
according to their relative strength. *Acids
have a value of less than 7 on the pH scale,
and bases have a value of more than 7.*

phy·lum (fī′ ləm) *noun* In classification,
a subgroup of a kingdom. *All animals
with a backbone are in the same phylum.*
phyla *plural*

phys·i·cal change (fiz′ i kəl chānj)
noun A change of matter from one form
to another without becoming a different
substance. *Water freezing to ice is a kind of
physical change.*

physical prop·er·ty (fiz′ i kəl
prop′ ər tē) *noun* A characteristic of
something that can be observed or
measured. *Weight and size are physical
properties.*

pis·til (pis′ təl) *noun* The female part of
a flower, where the seeds grow. *The pistil
is usually in the center of the flower.*

pitch (pich) *noun* How high or low a
sound is. *A foghorn has a low pitch.*

plain (plān) *noun* A large, flat area of land. *Buffalo once roamed the broad plains of the central United States.*

plan·et (plan' it) *noun* A large body that travels in an orbit around a star. *Our solar system contains nine planets.*

plant (plant) *noun* An organism that has many cells with thick cell walls, makes its own food, and is not able to move from place to place. *Plants, such as trees and flowers, make up one of the five kingdoms of living things.*

plas·ma (plaz' mə) *noun* The liquid part of blood. *Red and white blood cells float in plasma.*

plate (plāt) *noun* One of the very large sheets of rock that make up Earth's crust. *The plates float on the molten rock of Earth's mantle.*

pla·teau (pla tō') *noun* A broad area of high, flat land. *A plateau looks like a hill with a flat top.*

plate·let (plāt' lət) *noun* A tiny, flat object in blood that helps blood to clot. *Platelets help to stop a cut from bleeding.*

Plu·to (plü' tō) *noun* The farthest planet from the Sun and the smallest planet in our solar system. *Pluto appears to be a small, rocky planet covered with ice.*

po·lar cli·mate (pō' lər klī' mət) *noun* The very cold, windy, dry weather found all year round near Earth's North and South Poles. *The emperor penguins of Antarctica are well adapted to a polar climate.*

pole (pōl) *noun* One of the ends of a magnet. *A magnet's pull is strongest at its poles.*

pol·len (pol' ən) *noun* The yellow powder found in flowers that is needed to make seeds. *Bees help plants by carrying pollen.*

pol·lin·a·tion (pol i nā' shən) *noun* The transfer of pollen from the stamen of one flower to the pistil of another flower. *Pollination takes place when insects, birds, or the wind carry pollen from flower to flower.* **pollinate** *verb* **pollinator** *noun*

pol·lut·ant (pə lüt' ənt) *noun* A harmful substance that gets into the air, soil, or water. *Waste materials from factories can be pollutants in rivers.*

pol·lute (pə lüt') *verb* To make dirty. *Car exhaust pollutes the air around big cities.*

a	cat	e	net	îr	gear	u	cup	u̇	look, pull	th	this	ə	alive,
ā	day, lake	ē	seed	o	hot	ū	fuse	oi	soil	hw	wheel		comet,
ä	father	i	fit	ō	cold	ûr	fur, bird	ou	out	zh	measure		acid, atom,
âr	dare	ī	pine	ô	paw	ü	tool, rule	th	thin	ŋ	wing		focus

pol·lu·tion (pə lü′ shən) *noun*
The process of putting harmful substances in the air, soil, or water. *Water pollution can harm the fish in rivers.*

pop·u·la·tion (pop yə lā′ shən) *noun*
The total number of organisms of one type living in an area. *The world's population of blue whales is beginning to grow.*

pore (pôr) *noun* A gap or opening in a surface. *People sweat through pores in their skin.* **porous** *adjective*

pore space (pôr spās) *noun* A gap between soil particles or sand grains. *Pore spaces usually fill with water and air.*

po·ros·i·ty (pôr os′ ə tē) *noun*
A measure of the pore spaces in a rock or soil sample. *Porosity tells how much water soil can hold.*

po·si·tion (pə zish′ ən) *noun* The location of an object. *Earth's position in the solar system is third from the Sun.*

po·ten·tial en·er·gy (pō tən′ chəl en′ ər jē) *noun* The stored energy an object has because of its position. *A boulder at the edge of a cliff has potential energy.*

pound (pound) 1. *noun* A unit of weight. *The new baby weighs 8 pounds.* 2. *verb* To beat or throb. *Your heart pounds after a fast run.*

Pre·cam·bri·an er·a (prē kam′ brē ən îr′ ə) *noun* The earliest period of Earth's history, before the Paleozoic era. *Organisms with one cell appeared on Earth during the Precambrian era.*

pre·cip·i·ta·tion (pri sip i ta′ shən) *noun* Water in the atmosphere that falls to Earth. *Rain, snow, sleet, and hail are forms of precipitation.*

pred·a·tor (pred′ ə tər) *noun* An animal that hunts and eats other animals. *Lions, sharks, and eagles are predators.* **predatory** *adjective*

pre·dict (prə dikt′) *verb* To say what you think will happen. *What do you predict will happen if you put the plant in a dark room?* **prediction** *noun*

pre·scrip·tion (prə skrip′ shən) *noun* An order for drugs or medicine written by a doctor to a pharmacist. *The prescription tells how often to take your medicine.* **prescribe** *verb*

pres·er·va·tion (prez ər vā′ shən) *noun* The protection of an area as it is, without making changes to it. *One goal of the National Park Service is the preservation of wilderness lands.* **preserve** *verb*

pre·vail·ing winds (prē vā′ liŋ windz) *noun, plural* Winds that usually blow in the same direction. *The prevailing winds in North America blow from west to east.*

prey (prā) *noun* An animal hunted and eaten by another animal. *Birds and mice are the prey of cats.*

pri·mar·y con·su·mer (prī′ mâr ē kən sü′ mər) *noun* An animal that eats green plants. *Deer, grasshoppers, and rabbits are primary consumers.*

prism (priz′ əm) *noun* A thick piece of clear glass or plastic that can bend and reflect light. *A prism breaks up light into the colors of the rainbow.*

pro·bos·cis (prō bos′ is) *noun*
1. The long, tube-like mouthpiece of some insects. *A butterfly uses its proboscis like a straw to get nectar from a flower.*
2. A long nose that can bend. *An elephant's trunk is a proboscis.*

pro·du·cer (prə dü′ sər) *noun* An organism that makes food. *Green plants are producers.* **produce** *verb*

prop·er·ty (prop′ ər te) *noun* A characteristic of an object or substance that can be observed. *Color, shape, and texture are properties.*

pro·tec·tive re·sem·blance (prə tek′ tiv ri zem′ bləns) *noun* A type of camouflage in which an organism looks like an object in its environment. *The white fur of a polar bear against the snow is an example of protective resemblance.*

pro·tein (prō′ tēn) *noun* A substance necessary for life that is found in all living cells. *People get protein from foods such as eggs, milk, meat, and beans.*

pro·tist (prō′ tist) *noun* A tiny one-celled organism with a nucleus, such as some algae. *Protists make up one of the five kingdoms of living things.*

pro·ton (prō′ ton) *noun* One of the tiny particles in the nucleus of an atom. *A proton has a positive electric charge.*

pul·ley (pùl′ ē) *noun* A wheel with a groove in the rim, in which a rope or chain runs. *A pulley is a kind of simple machine.*

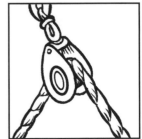

pulse (puls) *noun* The steady beat of the arteries as the heart pumps blood through them. *You can feel your pulse if you hold your wrist.*

pu·pa (pū′ pə) *noun* The third stage in the life cycle of some insects. *The pupa stage follows the larva stage.* **pupate** *verb*

pu·pil (pū′ pəl) *noun* The opening in the iris that lets light into the eye. *The pupil gets larger in dim light and smaller in bright light.*

a	cat	e	net	îr	gear	u	cup	ù	look, pull	th	this	ə	alive,
ā	day, lake	ē	seed	o	hot	ū	fuse	oi	soil	hw	wheel		comet,
ä	father	i	fit	ō	cold	ûr	fur, bird	ou	out	zh	measure		acid, atom,
âr	dare	ī	pine	ô	paw	ü	tool, rule	th	thin	ŋ	wing		focus

quartz (kwôrtz) *noun* A hard mineral found in many kinds of rocks. *Quartz comes in many forms and colors, from clear crystals to bright gemstones.*

queen bee (kwēn bē) *noun* The female bee that lays all the eggs for a colony. *A beehive usually has only one queen bee.*

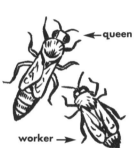

← queen

worker →

ra·di·al sym·me·try (rā′ dē əl sim′ ə trē) *noun* Having many sides that match up as mirror images. *Starfish have radial symmetry.*

ra·di·ate (rā′ dē āt) *verb* To send out energy in waves or rays. *The Sun radiates heat and light throughout the solar system.*

ra·di·a·tion (rā de ā′ shən) *noun* The transfer of energy through space. *Light, heat, X-rays, and microwaves are forms of radiation.*

ra·di·o·ac·tive el·e·ment (rā dē ō ak′ tiv el′ ə mənt) *noun* An element that gives off energy when the nuclei of its atoms give off particles. *Radium and uranium are examples of radioactive elements.*

radioactive waste (rā dē ō ak′ tiv wāst) *noun* Leftover material from nuclear energy plants. *Radioactive wastes are harmful to the environment.*

rain·bow (rān′ bō) *noun* A curved band of colors across the sky. *You can sometimes see a rainbow in the sky after a rainstorm.*

rain gauge (rān gāj) *noun* A device used to measure how much precipitation has fallen. *There was an inch of water in the rain gauge after the storm.*

range (rānj) *noun* The difference between the highest and lowest values in a set of numbers. *In the number set 10, 14, 20, 36, and 50, the range is 40.*

re·cep·tor (ri sep′ tər) *noun* A nerve ending that picks up information from the environment. *Receptors in the skin can sense pain, pressure, heat, and cold.*

re·charge·a·ble (rē chärj′ ə bəl) *adjective* Able to be used again and again. *The flashlight has a rechargeable battery.*

rec·la·ma·tion (rek lə mā′ shən) *noun* The repairing of damage done to an ecosystem. *After a forest fire, reclamation may include replanting trees.* **reclaim** *verb*

red blood cell (red blud sel) *noun* A cell in the blood that carries oxygen to the body's tissues and carries away carbon dioxide. *Red blood cells are produced in red bone marrow.*

re·duce (ri düs′) *verb* To make less or smaller. *It is important to reduce air pollution.* **reduction** *noun*

re·flect (ri flekt′) *verb* To bounce or throw back. *A mirror reflects light rays.*

re·flec·tion (ri flek′ shən) *noun* The bouncing of heat, light, or sound off a surface. *An echo is caused by the reflection of sound.*

re·flex (rē′ fleks) *noun* An automatic reaction of the body to something that happens to it. *Blinking your eyes at a flash of light is an example of a reflex.*

re·frac·tion (ri frak′ shən) *noun* The bending of light rays as they pass through different materials. *A telescope uses refraction through a lens to focus the view of a distant planet.* **refract** *verb*

re·gen·er·a·tion (rē jen ər ā′ shən) *noun* The growth of new body parts to replace ones that are lost or damaged. *The lizard grew a new tail by regeneration.* **regenerate** *verb*

rel·a·tive age (rel′ ə tiv āj) *noun* The age of something compared with the age of something else. *The relative age of the fossil found at the surface is younger than that of the fossil found 3 feet beneath it.*

relative hu·mid·i·ty (rel′ ə tiv hū mid′ i tē) *noun* The amount of water vapor in the air compared with the most that is possible at a certain temperature and air pressure. *The chance of rain increases as relative humidity nears 100 percent.*

relative mo·tion (rel′ ə tiv mō′ shən) *noun* Motion that is described by comparing one object's position to another object. *A model showed the relative motion of planets around the Sun.*

re·lo·cate (rē lō′ kāt) *verb* To find a new home. *Park rangers sometimes relocate bears away from campground areas.*

re·new·a·ble re·source (rē nü′ ə bəl re′ sôrs) *noun* A resource that can be replaced or used over and over again. *Wind and sunlight are renewable resources.*

re·pel (ri pel′) *verb* To push away. *The north poles of two magnets will repel each other.*

a	cat	e	net	îr	gear	u	cup	u̇	look, pull	*th*	this	ə	alive,
ā	day, lake	ē	seed	o	hot	ū	fuse	oi	soil	hw	wheel		comet,
ä	father	i	fit	ō	cold	ûr	fur, bird	ou	out	zh	measure		acid, atom,
âr	dare	ī	pine	ô	paw	ü	tool, rule	th	thin	ŋ	wing		focus

re·pro·duce (rē prə düs') *verb* To have offspring. *Living things reproduce and keep their species alive.*

re·pro·duc·tion (rē prə duk' shən) *noun* The way organisms make more of their own kind. *Plant reproduction usually needs either seeds or spores.*

rep·tile (rep' tīl) *noun* A cold-blooded animal that lives on land, lays eggs, and is covered with scales or plates. *Snakes, lizards, and crocodiles are reptiles.*

res·er·voir (rez' ər vwôr) *noun* A storage area for fresh water. *The water level in the reservoir got very low during the drought.*

re·sis·tance (ri zis' təns) *noun* The ability of a substance to oppose the flow of electric current through it. *Light bulb filaments heat up and glow because they have high resistance.*

re·sis·tor (ri zis' tər) *verb* A device that changes the amount of electric current that flows through a circuit. *A dimmer switch for a light uses a kind of resistor.*

re·source (rē' sôrs) *noun* Something found in nature that is valuable or useful. *Fresh water is an important natural resource.*

res·pir·a·tion (res pər ā' shən) *noun* 1. The release of energy from food within the cells of an organism. 2. The act of breathing. *Respiration involves inhaling and exhaling air.*

res·pi·ra·to·ry sys·tem (res' pər ə tôr ē sis' təm) *noun* The organ system that brings oxygen into the body and removes waste gas from the body. *The lungs are the main organs of the respiratory system.*

re·spond (ri spond') *verb* To react to conditions or changes in the environment. *Most plants respond to sunlight by growing toward it.*

re·sponse (ri spons') *noun* The way an organism acts when something happens. *Fear is a natural response to danger.*

ret·i·na (ret' ən ə) *noun* A layer of cells covering the back of the eye. *Light passes through the eye's lens and focuses on the retina.*

re·use (rē ūs') *verb* To use again instead of throwing away. *Reusing objects is a kind of recycling.* **reusable** *adjective*

rev·o·lu·tion (rev ə lü' shən) *noun* The movement of one object in a complete path around another object. *It takes Earth about 365 days to make one revolution around the Sun.*

re·volve (ri vôlv') *verb* To move in a circle or orbit around something. *The Moon revolves around Earth about every 28 days.*

Rich·ter scale (rik' tər skāl) *noun*
A measure of the strength of an
earthquake. *The earthquake measured 5.3
on the Richter scale.*

ridge (rij) *noun* A long, narrow chain of
mountains or hills. *The ridge blocked the
ocean from the plain.*

riv·er sys·tem (riv' ər sis' təm) *noun*
A major river and all the smaller rivers
and streams that flow into it. *The Ohio
and Missouri Rivers are part of the
Mississippi River system.*

rock (rok) *noun* The hard matter made
of minerals that forms Earth's crust. *Slate,
granite, and limestone are kinds of rock.*

rock cy·cle (rok sī' kəl) *noun*
An ongoing process by which rocks are
changed from one form to another.
*Weathering and erosion are part of the rock
cycle.*

rock de·bris (rok də brē') *noun* Gravel,
sand, and rock pieces that are picked up
by a glacier as it moves. *Even huge
boulders can be rock debris when carried by
a glacier.*

root (rüt) *noun*
The part of a plant
that holds the plant in
the ground. *Roots
absorb water and
minerals that the plant
needs from the soil.*

ro·tate (rō' tāt) *verb* To spin around a
center point or line (axis). *Earth rotates on
its axis once every 24 hours.*

ro·ta·tion (rō tā' shən) *noun*
The motion of a planet as it turns on
its axis. *The rotation of Earth causes night
and day.*

run·off (run' of) *noun* Water that flows
over Earth's surface but does not
evaporate or soak into the ground. *Runoff
after heavy rains can carry harmful
substances into rivers and lakes.*

sa·lin·i·ty (sə lin' ə tē) *noun*
The amount of salt in a solution. *Water
with high salinity is not safe to drink.*
saline *adjective*

sa·li·va (sə lī' və) *noun* The watery
liquid made in the mouth. *Saliva keeps
the mouth moist and helps to digest food.*

salt wa·ter (sôlt wô' tər) *noun* Water
with salt dissolved in it. *Oceans and seas
contain salt water.*

a	cat	e	net	îr	gear	u	cup	ủ	look, pull	*th* this	ə alive,
ā	day, lake	ē	seed	o	hot	ū	fuse	oi	soil	hw wheel	comet,
ä	father	i	fit	ō	cold	ûr	fur, bird	ou	out	zh measure	acid, atom,
âr	dare	ī	pine	ô	paw	ü	tool, rule	th	thin	ŋ wing	focus

sand dune (sand dün) *noun* A ridge or mound of sand piled up by wind. *Sand dunes can change shape and position over time.*

sat·el·lite (sat′ ə līt) *noun* An object that moves around another, larger object. *The Moon is a natural satellite of Earth.*

Sat·urn (sat′ ərn) *noun* The sixth planet from the Sun in our solar system. *Saturn is best known for its rings.*

sa·van·na (sə van′ ə) *noun* A flat, grassy plain with few or no trees. *Savannas are found in tropical areas such as central Africa.*

scale (skāl) *noun* 1. An instrument used to measure weight. *The scale shows that the stone weighs 2 pounds.* 2. The relationship between an object's actual size and the size of a drawing or model of it. *The scale of the model skeleton is 1 inch = 1 foot.* 3. A series of numbered marks that stand for units of measure. *The rain gauge has a centimeter scale.*

scav·en·ger (skav′ ən jər) *noun* An animal that feeds on dead animals and garbage. *Hyenas and vultures are scavengers.*

scent (sent) *noun* An odor given off by something. *A skunk uses its scent for defense against danger.*

sci·ence (sī′ əns) *noun* The study of nature and the physical world. *Science is based on observing and testing.*
scientific *adjective*

sci·en·tist (sī′ ən tist) *noun* An expert in one of the fields of science. *An astronomer is a scientist who studies objects in space.*

screw (skrü) *noun* A simple machine that is an inclined plane wrapped around a center post. *A screw looks like a nail with a spiral groove winding around it.*

sec·ond·ar·y con·su·mer (sek′ ən dâr ē kən sü′ mər) *noun* An animal that eats animals that eat green plants. *Wolves and hawks are secondary consumers.*

sed·i·ment (sed′ ə mənt) *noun* Particles that settle in a liquid. *A layer of sandy sediment covered the bottom of the tide pool.*

sed·i·men·ta·ry rock (sed ə mən′ tər ē rok) *noun* Rocks formed from layers of sediment pressed together. *Many fossils are found in sedimentary rocks.*

seed (sēd) *noun* The part of a plant that can develop and grow into a new plant. *Seeds contain stored food and are covered by a protective seed coat.*

seed coat (sēd kōt) *noun* The covering on the outside of a plant seed. *The seed coat protects the seed until it is ready to sprout.*

seedling ▶ sight

seed·ling (sēd' liŋ) *noun* A young plant grown from a seed. *The floor of the forest was covered with pine seedlings.*

seis·mic wave (sīz' mik wāv) *noun* A shaking caused by rocks moving at fault lines. *Earthquakes cause seismic waves.*

seis·mo·gram (sīz' mə gram) *noun* The record of seismic waves made by a seismograph. *The seismogram printout showed a mild earthquake.*

seis·mo·graph (sīz' mə graf) *noun* An instrument that records the strength of earthquake waves. *A seismograph can pick up very small ground motions.*

sense or·gans (sens ôr' gənz) *noun, plural* The body parts through which people become aware of the world around them. *The eyes, ears, nose, skin, and tongue are sense organs.*

se·pal (sē' pəl) *noun* A leaflike part that forms a ring around the base of a flower. *The sepals protect the flower bud before it blooms and protect the petals afterward.*

sep·tic tank (sep' tik taŋk) *noun* An underground storage tank where wastes are broken down by bacteria. *The waste and water from the drains of some houses are piped to a septic tank.*

se·ries cir·cuit (sîr' ēz sûr' kit) *noun* A circuit in which all the parts are connected one after the other. *The electric current in a series circuit follows one path.*

sew·age (sü' ij) *noun* Water mixed with human waste. *Most towns and cities have systems for treating and getting rid of sewage.*

sew·er (sü' ər) *noun* Large pipes or drains that carry sewage to treatment plants. *Sewers can overflow during heavy rains.*

shel·ter (shel' tər) *noun* A place where an animal or plant is protected from weather or danger. *Nests, burrows, and caves are types of shelter for animals.*

shoot·ing star (shüt' iŋ stär) *noun* Another name for a meteor. *Hundreds of shooting stars were seen during the meteor shower.*

short cir·cuit (shôrt sûr' kit) *noun* A circuit with too much electric current flowing through it. *Short circuits can overheat wires and cause fires.*

side ef·fect (sīd ə fekt') *noun* An unwanted result of taking medication. *A side effect of some allergy medicines is feeling sleepy.*

sight (sīt) *noun* The ability to see. *The sense organs for sight are the eyes.*

a	cat	e	net	îr	gear	u	cup	u̇	look, pull	th	this	ə	alive,
ā	day, lake	ē	seed	o	hot	ū	fuse	oi	soil	hw	wheel		comet,
ä	father	i	fit	ō	cold	ûr	fur, bird	ou	out	zh	measure		acid, atom,
âr	dare	ī	pine	ô	paw	ü	tool, rule	th	thin	ŋ	wing		focus

silt (silt) *noun* Tiny particles of rock that are smaller than sand and larger than clay. *Silt often settles as sediment in river deltas.*

sil·ver (sil′ vər) *noun* An element that is a white soft metal and is easy to shape. *Silver is often used for jewelry.*

sim·ple ma·chine (sim′ pəl mə shēn′) *noun* A tool with one or a few moving parts that makes work easier. *The six simple machines are the lever, wheel and axle, pulley, inclined plane, wedge, and screw.*

skel·e·tal mus·cle (skel′ i təl mus′ əl) *noun* A muscle that is attached to a bone and allows movement. *Arm and leg muscles are skeletal muscles.*

skeletal sys·tem (skel′ i təl sis′ təm) *noun* The body system that gives the body shape and structure and protects organs. *The skeletal system is made up of bones, cartilage, and ligaments.*

skel·e·ton
(skel′ i tən) *noun* The framework of bones in an animal's body. *The bones in a bird's skeleton are very light.*

small in·tes·tine
(smôl in tes′ tən) *noun* The long tube that food passes through after it leaves the stomach. *Most digestion takes place in the small intestine.*

smog (smog) *noun* A mixture of smoke and fog. *Smog often pollutes the air around large cities.*

smooth mus·cle (smŭ*th* mus′ əl) *noun* A type of muscle in the walls of some organs and blood vessels. *Smooth muscle helps move food through the digestive system.*

so·cial an·i·mal (sō′ shəl an′ ə məl) *noun* An animal that lives in groups rather than on its own. *Ants, bees, and even elephants are social animals.*

soil (soil) *noun* The layer of loose material covering Earth's land surface. *Plants grow well in our garden soil.*

soil con·ser·va·tion (soil kon sûr vā′ shən) *noun* Practices that protect the soil and its quality. *Soil conservation allows soil to be used for planting again and again.*

soil pro·file (soil prō′ fīl) *noun* A vertical section of soil from surface down to bedrock. *The layers in a soil profile are called horizons.*

soil wa·ter (soil wô′ tər) *noun* Water that soaks into the ground. *Clay can hold a lot of soil water.*

so·lar e·clipse (sō′ lər ē klips′) *noun* The temporary blocking of the Sun's light as the Moon passes between Earth and the Sun. *We see only a ring of sunlight around the new Moon during a solar eclipse.*

solar en·er·gy (sō′ lər en′ ər jē) *noun* The energy given off by the Sun. *Some buildings are heated by solar energy.*

solar sys·tem (sō′ lər sis′ təm) *noun* A star and all the objects that move around it. *Our solar system contains nine planets.*

sol·id (sol′ əd) *noun* A form of matter that has a set shape and a set volume. *Wood and metal are solids.* **solid** *adjective*

sol·u·bil·i·ty (sol yə bil′ ə tē) *noun* A measure of the amount of substance that will dissolve in another substance. *Sugar and salt have high solubility in water.* **soluble** *adjective*

sol·u·tion (sə lü′ shən) *noun* A mixture formed by dissolving one substance in another. *Hot chocolate is a solution of chocolate powder and hot milk.*

son·ic boom (son′ ik büm) *noun* A loud noise heard when a jet plane flies faster than the speed of sound. *Sonic booms sound like explosions.*

sound (sound) *noun* A series of vibrations that can be heard. *The ear is the sense organ that picks up sound.*

sound wave (sound wāv) *noun* A vibration in air or water caused by a moving object. *A drum and a guitar make different kinds of sound waves.*

South·ern Hem·i·sphere (su*th*′ ərn hem′ i sfîr) *noun* The half of Earth that is south of the equator. *Australia is in the Southern Hemisphere.*

South Pole (south pōl) *noun* The most southern point on Earth. *The South Pole is on the continent of Antarctica.*

space (spās) *noun* 1. The region that stretches in all directions, has no limits, and contains everything in the universe. *We use telescopes to view objects in outer space.* 2. The area inside or around an object. *A magnetic field exists in the space around a magnet.*

space probe (spās prōb) *noun* A remote-controlled space vehicle with cameras and scientific tools for exploration. *The Voyager 1 and 2 space probes were used to study the outer planets except for Pluto.*

spe·cies (spē′ shēz) *noun* The smallest classification group, made up of only one type of organism. *Every kind of plant and animal has its own species.*

spec·trum (spek′ trəm) *noun* The colors of light we can see. *Seven colors make up the spectrum: red, orange, yellow, green, blue, indigo, and violet.*

speed (spēd) *noun* A measure of how fast an object moves. *The bus traveled at a speed of 80 kilometers (50 miles) per hour.*

a	cat	e	net	îr	gear	u	cup	u̇	look, pull	*th*	this	ə	alive,
ā	day, lake	ē	seed	o	hot	ū	fuse	oi	soil	hw	wheel		comet,
ä	father	i	fit	ō	cold	ûr	fur, bird	ou	out	zh	measure		acid, atom,
âr	dare	ī	pine	ô	paw	ü	tool, rule	th	thin	ŋ	wing		focus

speed of light (spēd əv līt) *noun*
The rate at which a light ray travels in empty space. *The speed of light is 299,792 kilometers (186,282 miles) per second.*

speed of sound (spēd əv sound) *noun*
The rate at which a sound wave travels. *Some airplanes fly faster than the speed of sound.*

spher·i·cal sym·me·try (sfîr' ə kəl sim' ə trē) *noun* Having a round ball shape that would match any way it was divided in half. *A globe has spherical symmetry.*

spi·nal cord (spi' nəl kôrd) *noun* The bundle of nerves that runs through the backbone. *The spinal cord carries signals between the brain and the rest of the body.*

sponge (spunj) *noun* A sea animal with a rubbery skeleton and many holes for soaking up water. *A sponge pumps water through its body to get food and oxygen.*

spore (spôr) *noun* A plant cell that can grow into a new plant. *Some plants, like ferns and mosses, produce spores.*

sprain (sprān) *verb* To injure a joint by twisting or tearing a muscle or ligament. *The runner sprained her ankle when she tripped on the curb.*

spring scale (spriŋ skāl) *noun* A tool used to measure the force of gravity on an object. *A spring scale gives a reading in newtons.*

sta·ble (stā' bəl) *noun* Steady or not easily changed. *Our body temperature remains stable at about 37°C (98.6°F).* **stability** *noun*

sta·men (stā' mən) *noun* The part of a flower that produces pollen. *The stamens surround the pistil in the middle of a flower.*

stand·ard u·nit (stan' dərd ū' nit) *noun* A unit of measure that is always the same and that all people who use it agree on. *Meters and liters are standard units in the metric system.*

star (stär) *noun* A huge ball of glowing gases in space. *The Sun is the nearest star to Earth.*

state of mat·ter (stāt əv mat' ər) *noun* The physical form of a substance. *Solid, liquid, and gas are states of matter.*

stat·ic elec·tric·i·ty (stat' ik i lek tris' i tē) *noun* An electric charge that builds up on an object and stays there. *You can make static electricity by rubbing a balloon against your sweater.*

stem (stem) *noun* The main part of a plant that grows up from the ground. *The stem holds the plant up and carries water up from the roots.*

steth·o·scope (steth' ə skōp) *noun* A medical instrument used to listen to the body's sounds. *Doctors use stethoscopes to hear how well the heart and lungs are working.*

stim·u·lant (stim' yə lənt) *noun* A substance that speeds up the activity of the body. *Coffee and some soft drinks contain stimulants.* **stimulate** *verb*

stim·u·lus (stim' yə ləs) *noun*
Something that causes action or activity.
The stimulus of touching a hot pan caused the boy to pull away his hand.
stimuli *plural*

stom·ach (stum' ək) *noun* The organ where food goes after it is swallowed. *Food is partly digested by juices in the stomach.*

storm surge (stôrm sûrj) *noun* A very large series of ocean waves caused by a hurricane. *The storm surge flooded coastal towns.*

strat·o·sphere (strat' ə sfîr) *noun* The layer of atmosphere from about 15–50 kilometers (about 10–30 miles) above Earth, where the air is thin and cold. *Some jet planes fly in the stratosphere because it is above Earth's weather.*

strat·us cloud (strat' əs kloud) *noun* A low gray cloud that covers a very large area. *Stratus clouds often bring light rain or snow.*

streak (strēk) *noun* A long, thin mark or stripe made by rubbing a mineral against a tile surface. *Streak color can help identify a mineral.*

streak plate (strēk plāt) *noun* A tile plate that a mineral is rubbed against to see what color streak it makes. *Graphite leaves a black streak on a streak plate.*

stride (strīd) *noun* The distance traveled in a single step. *An adult's stride is longer than a child's stride.* **stride** *verb*

strip crop·ping (strip krop' iŋ) *noun* A kind of farming in which strips of one crop are planted between strips of another crop. *Strip cropping is a way to prevent soil erosion.*

sub·soil (sub' soil) *noun* A hard layer of soil just below the topsoil. *Subsoil is usually made up of clay and minerals.*

sub·stance (sub' stəns) *noun* Matter, or anything that has weight and takes up space. *Liquids, powders, and solid objects are types of substances.*

suc·ces·sion (sək sesh' ən) *noun* A gradual process by which an ecosystem or a species changes over time. *The swamp changed into a forest through succession.* **successive** *adjective*

Sun (sun) *noun* The star that is the center of our solar system. *The Sun has been burning for about 5 billion years.*

sun·spot (sun' spot) *noun* A dark patch on the surface of the Sun. *Sunspots are caused by magnetic storms that cool a large area.*

a	cat	e	net	îr	gear	u	cup	u̇	look, pull	th	this	ə	alive,
ā	day, lake	ē	seed	o	hot	ū	fuse	oi	soil	hw	wheel		comet,
ä	father	i	fit	ō	cold	ûr	fur, bird	ou	out	zh	measure		acid, atom,
âr	dare	ī	pine	ô	paw	ü	tool, rule	th	thin	ŋ	wing		focus

sur·face cur·rent (sûr' fəs kûr' ənt) *noun* An ocean current caused by steady winds blowing over the surface of the water. *Surface currents and density currents are two kinds of ocean currents.*

surface ten·sion (sûr' fəs ten' shən) *noun* The pulling together of the layer of particles at the surface of a liquid. *We filled glasses with water to study surface tension.*

switch (swich) *noun* A device that opens and closes an electric circuit. *This wall switch controls the lights in the room.* **switch** *verb*

sym·bi·o·sis (sim bē ō' sis) *noun* Two different species of organism living together. *The pilot fish and the shark are an example of symbiosis.*

sym·me·try (sim' ə trē) *noun* A matching design or setup of parts on both sides of a center line. *The human body has symmetry.* **symmetrical** *adjective*

sys·tem (sis' təm) *noun* A group of things or parts that work together. *The body is made up of several organ systems.*

tap·root (tap' root) *noun* A large, thick root that grows straight down below a plant. *Taproots can grow deep into the soil.*

tar (tär) *noun* 1. A thick, dark, sticky substance found in cigarettes. *Tar has been proven to be a cause of cancer.* 2. A thick, black, sticky substance made from wood or coal. *Tar is used to cover roofs and road surfaces.*

taste buds (tāst budz) *noun, plural* Clusters of cells on the tongue that sense flavors. *Our taste buds tell us whether a food is sweet, sour, salty, or bitter.*

tel·e·scope (tel' ə skōp) *noun* An instrument that uses lenses and sometimes mirrors to make distant objects look larger and closer. *Telescopes are often used for observing objects in space.*

tem·per·ate cli·mate (tem' pər ət klī' mət) *noun* A climate that is neither very hot nor very cold. *Most of the United States has a temperate climate.*

tem·per·a·ture (tem' pər ə chər) *noun* A measure of how hot or cold something is. *We keep the temperature in our house around 20°C (68°F).*

ten·don (ten' dən) *noun* A strong band of tissue that connects a muscle to a bone. *The pitcher had a sore shoulder tendon after the ball game.*

ter·mi·nal (tûr' mə nəl) *noun* A place where a wire can be connected to a battery. *The battery has a positive terminal and a negative terminal.*

ter·mi·nus (tûr' mə nəs) *noun* The outer edge of a glacier. *Rock debris is piled up at the glacial terminus.*

ter·rac·ing (ter' is iŋ) *noun* A kind of farming in which crops are planted on stepped levels up a slope. *Terracing is often used in hilly areas.* **terrace** *verb, noun*

therm·al (thûr' məl) *adjective* Having to do with heat. *Old Faithful in Yellowstone National Park is a thermal geyser.*

ther·mom·e·ter (thûr mom' ə tər) *noun* An instrument used to measure temperature. *We check the thermometer twice a day for our weather project.*

thor·ax (thôr' ax) *noun* The center part of an insect's three body parts. *An insect's legs are attached to its thorax.*

thun·der (thun' dər) *noun* The loud noise that follows a flash of lightning. *The thunder was a warning that a storm was coming.* **thunder** *verb*

thun·der·head (thun' dər hed) *noun* A large cloud charged with static electricity that produces thunder and lightning. *Thunderheads are dark with flat bases and puffy tops.*

thun·der·storm (thun' dər stôrm) *noun* A rainstorm with thunder and lightning. *Thunderstorms often occur on hot, humid summer days.*

tide (tīd) *noun* The regular rise and fall of ocean levels, caused by the pull of the Sun and Moon on Earth. *High tides occur about every 12$\frac{1}{2}$ hours.* **tidal** *adjective*

tis·sue (tish' ü) *noun* A group of similar cells that work together to do a job. *Muscles are made of muscle tissue.*

top·soil (top' soil) *noun* The surface layer of soil, in which plants grow. *Topsoil is rich in minerals and humus.*

tor·na·do (tôr nā' dō) *noun* A powerful, whirling wind that damages or destroys most things in its narrow path. *Tornadoes are dark, funnel-shaped clouds.*

trace fos·sil (trās fos' əl) *noun* A fossil that shows a change that an animal made in its environment long ago. *A dinosaur track in a dry riverbed is an example of a trace fossil.*

trait (trāt) *noun* A characteristic of an organism. *A spotted coat and great speed are traits of the cheetah.*

tran·quil·i·zer (traŋ' kwə lī zər) *noun* A medicine used to calm a person. *The doctor gave a tranquilizer to the man who was in the accident.* **tranquil** *adjective* **tranquilize** *verb*

a	cat	e	net	îr	gear	u	cup	u̇	look, pull	*th* **this**	ə	alive,
ā	day, lake	ē	seed	o	hot	ū	fuse	oi	soil	hw **wheel**		comet,
ä	father	i	fit	ō	cold	ûr	fur, bird	ou	**out**	zh **measure**		acid, atom,
âr	dare	ī	pine	ô	paw	ü	tool, rule	th	thin	ŋ	wi**ng**		focus

trans·for·mer (trans fôr′ mər) *noun*
A piece of equipment that changes the voltage of an electric current. *We use transformers to control the strength of electricity.*

trans·lu·cent (trans lü′ sənt) *adjective*
Letting light pass through, but not clear. *Window glass is transparent, but frosted glass is translucent.*

trans·mit (trans mit′) *verb* To send or pass on from one place or person to another. *Traits are transmitted from parent to offspring through genes.*
transmission *noun*

trans·par·ent (trans per′ ənt) *adjective*
Letting most light pass through. *You can see clearly through a transparent substance.*

tran·spi·ra·tion (tran spə rā′ shən)
noun The giving off of water vapor by plants. *Transpiration takes place through tiny openings on the leaves.* **transpire** *verb*

trench (trench) *noun* A deep, narrow valley on the ocean floor where Earth's plates meet. *Trenches in the Pacific Ocean are some of the deepest places on Earth.*

tri·ceps (trī′ seps) *noun, singular, plural*
The muscle at the back of the upper arm. *The triceps straightens the arm by contracting, or getting shorter.*

tri·lo·bite (trī′ lə bīt)
noun A small, prehistoric sea animal, now extinct. *Trilobite fossils have been found in many parts of the world.*

trop·i·cal (trop′ i kəl) *adjective* Typically having very hot weather. *Areas near the equator have tropical climates.*

trop·i·cal rain for·est (trop′ i kəl rān fôr′ əst) *noun* A hot, wet forest with very tall trees. *The largest tropical rain forest in the world is in Brazil.*

tro·po·sphere (trō′ pə sfir) *noun*
The layer of atmosphere closest to Earth, about 10–16 kilometers (6–10 miles) thick. *Earth's weather occurs in the troposphere.*

trough (trof) *noun* The lowest part of a wave. *Wave height is the distance from the trough to the crest.*

true-form fos·sil (trü′ fôrm fos′ əl)
noun The fossilized remains of the actual animal or animal part. *An ant preserved in amber is a true-form fossil.*

tsu·na·mi (tsü nä′ mē) *noun* A huge wave that travels at high speed across the ocean. *Tsunamis are caused by underwater earthquakes or volcanoes.*

tu·ber (tü′ bər) *noun* A thick underground stem. *A potato is a tuber.*
tuberous *adjective*

tun·dra (tun′ drə) *noun* A cold, flat area with no trees. *Many parts of Alaska are tundra.*

u·ni·verse (ū′ ni vûrs) *noun* All space and everything in it. *The universe contains many hundreds of billions of stars.*

u·ra·ni·um-lead dat·ing meth·od (yu̇ rā′ ne əm led dā′ tiŋ meth′ əd) *noun* A way to date the age of rocks. *The uranium-lead dating method helped scientists learn how old Earth is.*

Ur·a·nus (yu̇r′ ə nəs) *noun* The seventh planet from the Sun in our solar system. *Uranus spins on its side instead of upright like the other planets.*

u·rine (yu̇r′ in) *noun* The liquid waste that animals pass out of their bodies. *Urine is stored in the bladder.*

vac·ci·na·tion (vak si nā′ shən) *noun* A protection against a disease by introducing dead or weakened germs into the body. *Many people get vaccinations when they are babies.*
vaccinate *verb*

vac·cine (vak sēn′) *noun* A substance that contains dead or weakened organisms and that can be injected or swallowed. *A vaccine makes the body produce substances that fight a disease.*

vac·u·ole (vak′ ū ōl) *noun* A tiny space in a cell. *Vacuoles are where cells hold food, water, and wastes.*

vac·u·um (vak′ ū əm) *noun* Empty space. *A vacuum contains no air or other matter.*

val·ley (val′ ē) *noun* A low area between hills or mountains. *A river often flows through a valley.*

var·i·a·ble (ver′ ē ə bəl) *noun* The factor being tested in an experiment. *The variable in our experiment was the type of soil we planted the seeds in.*

a	cat	e	net	îr	gear	u	cup	u̇	look, pull	th	this	ə	alive,
ā	day, lake	ē	seed	o	hot	ū	fuse	oi	soil	hw	wheel		comet,
ä	father	i	fit	ō	cold	ûr	fur, bird	ou	out	zh	measure		acid, atom,
âr	dare	ī	pine	ô	paw	ü	tool, rule	th	thin	ŋ	wing		focus

var·i·a·tion (ver ē ā' shən) *noun*
A small difference between two similar objects. *All snowflakes have six sides, but their patterns have variations.* **vary** *verb*

vas·cu·lar plant (vas' kyə lər plant) *noun* A plant with tubelike vessels for carrying water and food. *Trees are large vascular plants.*

veg·e·ta·ble (vej' tə bəl) *noun* A plant or plant part that is used as food. *Lettuce, celery, and broccoli are vegetables.*

vein (vān) *noun* A blood vessel that carries blood toward the heart. *Our veins sometimes look blue through our skin.*

vent (vent) *noun* An opening through which heat or heated material escapes. *Lava and gases rise through the vent in a volcano.*

ven·tri·cle (ven' trə kəl) *noun* One of the two lower chambers or parts of the heart. *The ventricles pump blood out to the lungs and to the rest of the body.*

Ve·nus (vē' nəs) *noun* The second planet from the Sun in our solar system. *Venus is sometimes called Earth's twin because it is almost the same size as our planet.*

ver·te·bra (vûr' tə brə) *noun* One of the bones that make up the backbone. *Our vertebrae help us stand up straight.* **vertebrae** *plural*

ver·te·brate (vûr' tə brāt) *noun* An animal with a backbone. *Amphibians, fish, birds, reptiles, and mammals are vertebrates.*

vi·brate (vī' brāt) *verb* To move back and forth rapidly. *A tuning fork vibrates when it is struck.* **vibration** *noun*

vi·rus (vī' rəs) *noun* A very tiny particle that can reproduce only inside living cells. *Colds and flu are caused by viruses.*

vi·ta·min (vīt' ə min) *noun* A substance needed by the body to keep healthy. *Orange juice is a good source of vitamin C.*

vol·can·ic moun·tain (vol kān' ik moun' tən) *noun* A mountain formed by a volcano. *Many islands in the Pacific Ocean are the tops of volcanic mountains.*

vol·ca·no (vol kā' nō) *noun* A mountain with vents through which lava, gases, and ash erupt. *Some volcanoes form where magma breaks through Earth's surface.*

volt (vōlt) *noun* A unit for measuring the force that is available to push an electric current in a circuit. *The smoke detector needs a 9-volt battery.*

volt·age (vōl' tij) *noun* The force that makes electric charges move in an electric current. *Transformers lower the voltage to levels that are safe to use in houses.*

vol·ume (vol' ūm) *noun* 1. The amount of space an object takes up. *The volume of water in the pitcher is 2 liters (about 2 quarts).* 2. How loud or soft a sound is. *A whisper has low volume, and a shout has high volume.*

vol·un·tar·y mus·cle (vol' ən ter ē mus' əl) *noun* A muscle controlled by a person's thinking. *The muscles that allow us to move are voluntary muscles.*

warm-blood·ed (wôrm' blud əd) *adjective* Having a body temperature that stays the same, no matter what the surrounding temperature. *Mammals and birds are warm-blooded animals.*

warm front (wôrm frunt) *noun* A boundary between air masses where warm air is taking the place of cold air. *The warm front brought wet weather.*

wa·ter (wô' tər) *noun* A clear liquid that falls as rain and fills oceans, rivers, and lakes. *Water is the most common substance on Earth's surface and in organisms.*

water con·ser·va·tion (wô' tər kon sûr vā' shən) *noun* Practices that protect water and reduce its use. *Water conservation protects an important resource from being wasted.*

water cy·cle (wô' tər sī' kəl) *noun* The continuous movement of water between the atmosphere and Earth's surface. *The water cycle includes evaporation, condensation, and precipitation.*

water pol·lu·tion (wô' tər pə lü' shən) *noun* Harmful materials that make water unclean or unsafe. *Some water pollution comes from sewage.*

water ta·ble (wô' tər tā' bəl) *noun* The upper area of groundwater. *A well must be deep enough to reach below the water table.*

water treat·ment plant (wô' tər trēt' mənt plant) *noun* A place where water is cleaned. *Our town has a water treatment plant to make our drinking water safe.*

water va·por (wô' tər vā' pər) *noun* Water in the form of a gas. *Water vapor is in the air, but we cannot see it.*

wave (wāv) *noun* 1. A vibration of energy. *Sound waves travel easily through air and water.* 2. A curving edge of water that moves along the surface of the ocean. *The up-and-down motion of a wave travels, but the water does not.*

a	cat	e	net	îr	gear	u	cup	ú	look, pull	th	this	ə	alive,
ā	day, lake	ē	seed	o	hot	ū	fuse	oi	soil	hw	wheel		comet,
ä	father	i	fit	ō	cold	ûr	fur, bird	ou	out	zh	measure		acid, atom,
âr	dare	ī	pine	ô	paw	ü	tool, rule	th	thin	ŋ	wing		focus

wave·length (wāv′ leŋ*th*) *noun*
The distance between the top of one wave and the top of the next wave. *A high-pitched sound has a short wavelength.*

weath·er (we*th*′ ər) *noun* The condition of the atmosphere at a given place and time. *Temperature, air pressure, and wind are important parts of weather.*

weather bal·loon (we*th*′ ər bə lün′) *noun* A balloon that carries weather instruments high in the atmosphere to measure temperature, humidity, and air pressure at different altitudes. *Weather balloons are launched twice every day from over 1,000 stations around the world.*

weather fore·cast (we*th*′ ər fôr′ kast) *noun* A prediction of what the weather will be, based on collected data. *The weather forecast calls for a clear, cool, breezy day tomorrow.*

weath·er·ing (we*th*′ ər iŋ) *noun* The breaking down of rocks into smaller pieces. *Ice, wind, rain, and plants cause weathering.*

weather map (we*th*′ ər map) *noun* A chart that shows weather conditions in a certain region. *Weather maps use special symbols to show different kinds of weather.*

weather sat·el·lite (we*th*′ ər sat′ ə līt) *noun* An object that orbits Earth above the atmosphere and keeps track of weather conditions. *Most weather satellites have cameras to take pictures of storms and clouds.*

wedge (wej) *noun* A simple machine made by combining two inclined planes back to back. *An axe blade is a wedge used to split wood.*

weigh (wā) *verb* To measure weight or the pull of gravity on an object. *We use a scale to weigh things.*

weight (wāt) *noun* The pull of gravity on an object. *The weight of all the schoolbooks made the backpacks hard to lift.*

wet cell (wet sel) *noun* A power source that produces electric current using two different metal bars in an acid solution. *Wet cell batteries are used in cars.*

wheel and ax·le (hwēl ənd ak′ səl) *noun* A simple machine made up of a large wheel attached at the center to a rod or handle. *A car's steering wheel and column are a wheel and axle.*

white blood cell (hwīt blud sel) *noun* A colorless blood cell that fights harmful germs that enter the body. *White blood cells are part of the body's immune system.*

wind (wind) *noun* The movement of air. *The wind is blowing hard from the north today.* **windy** *adjective*

wind en·er·gy (wind en′ ər jē) *noun* The energy of the movement of air. *Wind energy turns the blades of a windmill to produce electricity.*

wind·pipe (wind′ pīp) *noun* The tube that connects the back of the mouth with the lungs. *Air passes through the windpipe when you breathe.*

wind·sock (wind′ sok) *noun* A baglike object that fills with air when the wind blows into it. *Windsocks show the direction and strength of the wind.*

wind vane (wind vān) *noun* An instrument that spins freely to show the direction of the wind. *The arrow of a wind vane points in the direction the wind is blowing from.*

work (wûrk) *noun* The result of a force moving an object over a distance. *Lifting a book from the floor onto a table is work.*

work·er bee (wûrk′ ər bē) *noun* A female bee that does work for the colony but does not lay eggs. *Worker bees build hives and gather food.*

x-ax·is (eks′ ak sis) *noun* The scale of values along the base line of a graph. *The x-axis shows the years since 2000.*

X-ray (eks′ rā) *noun* An invisible ray of light that can pass through solid objects. *Doctors use X-rays to see organs and bones inside the body.*

y-ax·is (wī′ ak sis) *noun* The scale of values along a line up and down the side of a graph. *The y-axis shows the bald eagle population, in thousands.*

year (yîr) *noun* The time it takes a planet to orbit the Sun. *A year on Earth is about 365 days.*

young (yuŋ) 1. *noun* An offspring. *Bears take care of their young for up to 2 years after birth.* 2. *adjective* In an early part of life. *A cub is a young bear.*

a	cat	e	net	îr	gear	u	cup	u̇	look, pull	th	this	ə	alive,
ā	day, lake	ē	seed	o	hot	ū	fuse	oi	soil	hw	wheel		comet,
ä	father	i	fit	ō	cold	ûr	fur, bird	ou	out	zh	measure		acid, atom,
âr	dare	ī	pine	ô	paw	ü	tool, rule	th	thin	ŋ	wing		focus